Don't Mess with Maths

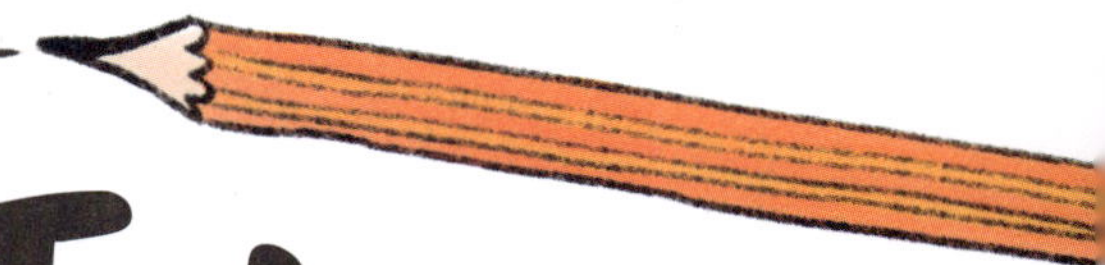

BY ANNA CLAYBOURNE

ILLUSTRATED BY SHANARAMA

ARCTURUS

ARCTURUS

This edition published in 2023 by Arcturus Publishing Limited
26/27 Bickels Yard, 151–153 Bermondsey Street,
London SE1 3HA

Author: Anna Claybourne
Illustrator: Shanarama
Stairs artwork on page 64: Shutterstock
Designer: Jeni Child
Editor: Violet Peto
Design Manager: Jessica Holliland
Editorial Manager: Joe Harris

Published and Distributed in India by Scholastic India Pvt. Ltd.
This reprint edition: February, 2025

ISBN: 978-93-5954-515-8

Printed in India at Polykam offset New Delhi 110028

Contents

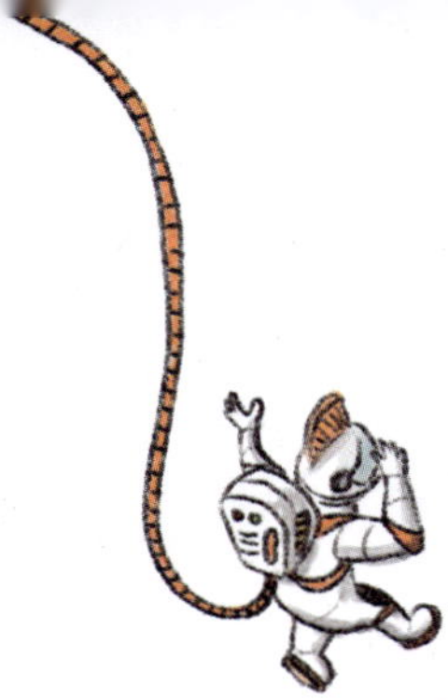

Explore, experiment, get messy!

This book is about all the weird, crazy, and fun things you can do with numbers, shapes, patterns, and all kinds of messy mathematical stuff!

In these pages, we'll show you how to ...

- Draw, make, measure, model, and create!

- Explore 3D shapes, cool codes, curious curves, natural numbers, and mind-blowing mazes ...

- Discover genius number tricks, shortcuts, and surprises ...

Height = arm span!

• Challenge your friends and family to solve baffling puzzles ...

• Play perplexing tricks on your own brain ...

• And draw and make impossible things!

Mess it up!

Most of all, this book is about experimenting, trying things out, and seeing what you can discover. On every page, there's space to try out a mathematical puzzle, art creation, challenge, or experiment. And when you've done that, you can keep experimenting on your own.

Turn the page, and let the mathematical messiness begin ...

Around in Circles

What is a circle, and what can you do with it?

All the way around

In a circle, the edge is always the same distance from the middle—all the way around.

That's all there is to it!

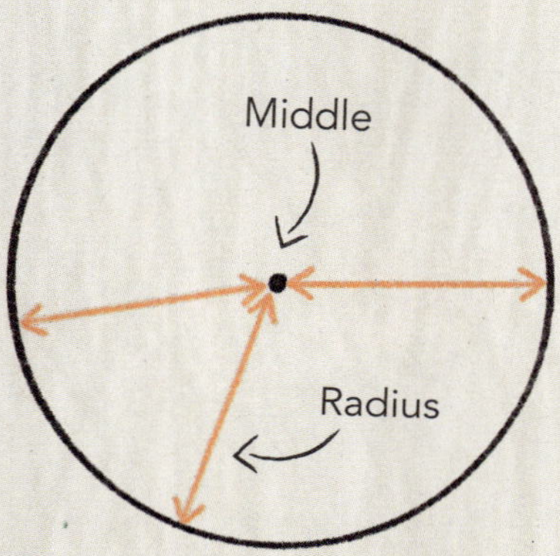

This distance is called the radius.

Draw a circle

There are a few ways to draw a great circle …

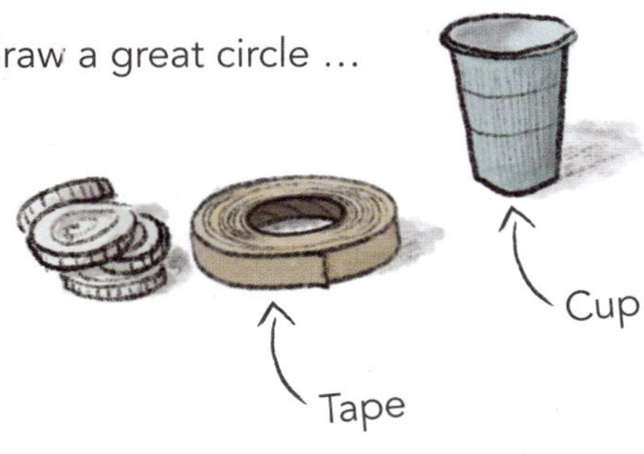

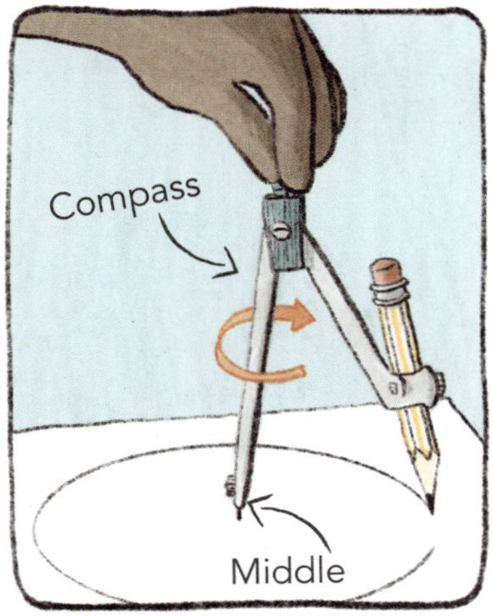

A compass holds a pen or pencil a fixed distance away from a point, so it draws a circle!

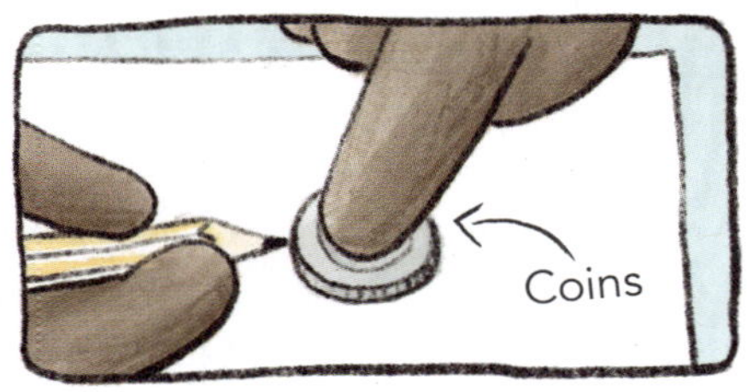

Or just draw around something round!

CLEVER TRICK!

If you don't have a compass or a round object, here's a clever circle-drawing trick.

Draw a dot on some paper.

Put your little finger on the dot.

Hold a pencil in the same hand, with the tip on the paper.

Hold your hand still, and with your other hand, turn the paper around (or ask someone else to do this).

Circle art

Now use your circles!

Seven circles together make a flower.

Overlapping circles
make a flower pattern.

For a perfect heart,
draw a square …

add circles here …

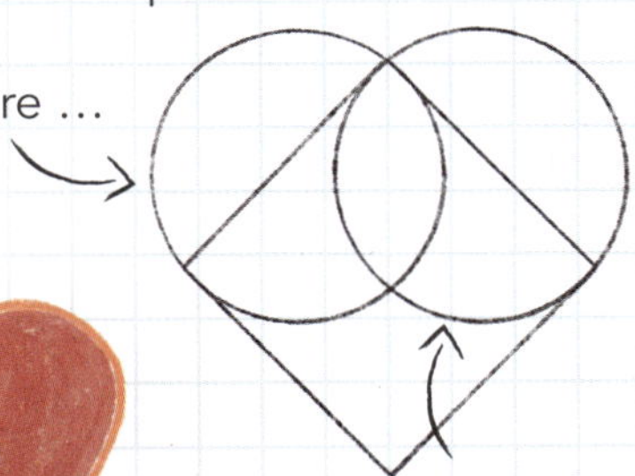

then erase
these lines

Use a row of circles
to make a caterpillar!

Space to draw in

Wheely cool!

The shape of a circle is perfect for making wheels.

Curious triangle

So are there any other shapes like this?

Meet a very special shape … the Reuleaux triangle!

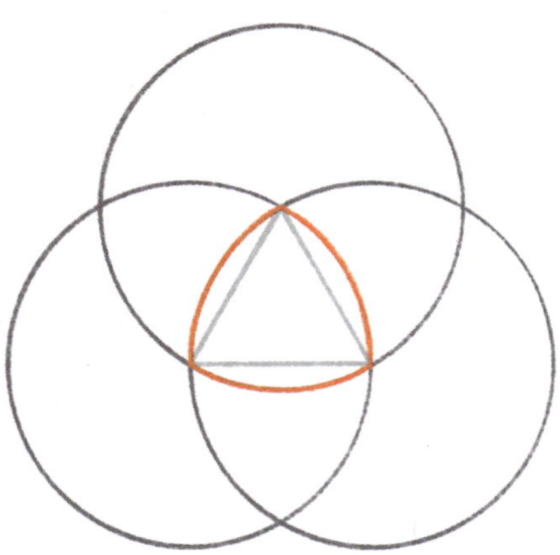

In a circle, the diameter (distance across the middle) is always the same.

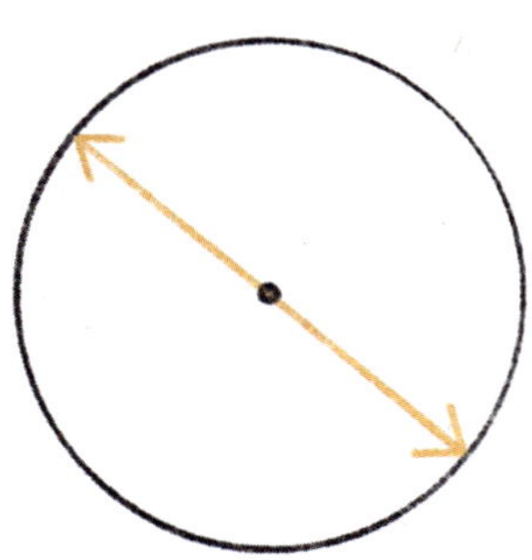

And the same goes for a Reuleaux triangle.

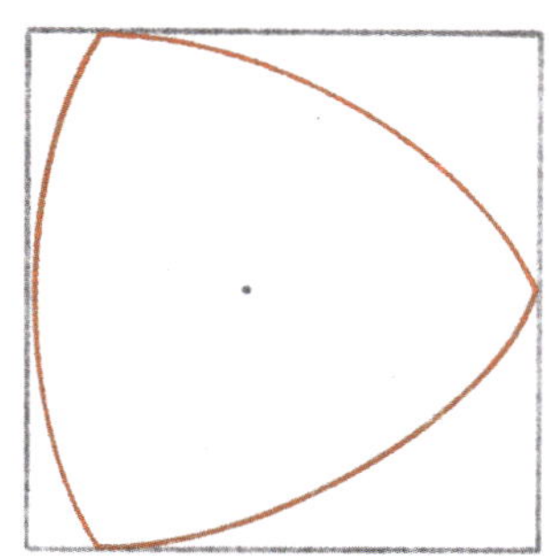

This does NOT work with a normal triangle. Try measuring them across the dot in the middle and see!

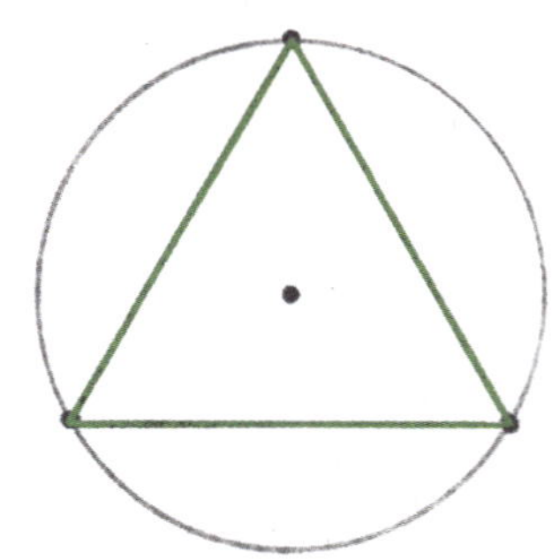

Triangle wheels

So do Reuleaux triangles make good wheels?

To find out, trace the shapes on this page onto thick card, and make two round wheels and two Reuleaux triangle wheels.

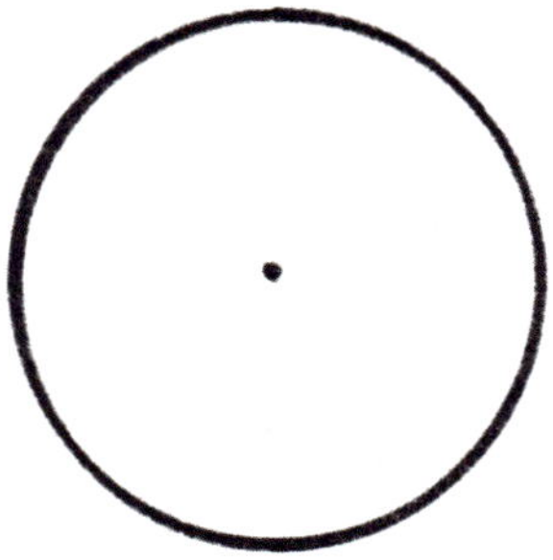

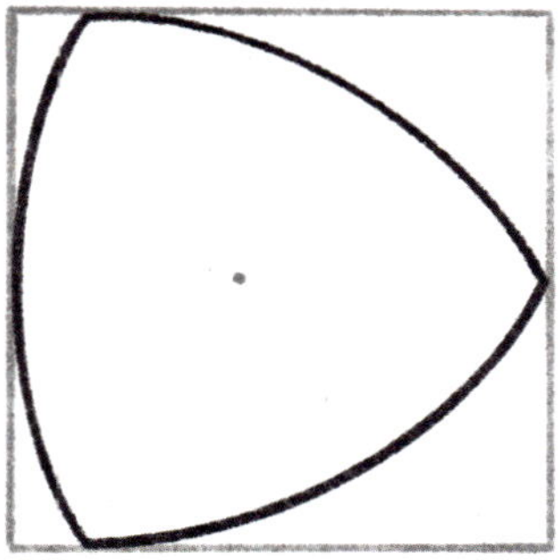

Use wooden skewers or cocktail sticks as axles …

Then roll your wheels along with a thin, flat book on top. What happens?

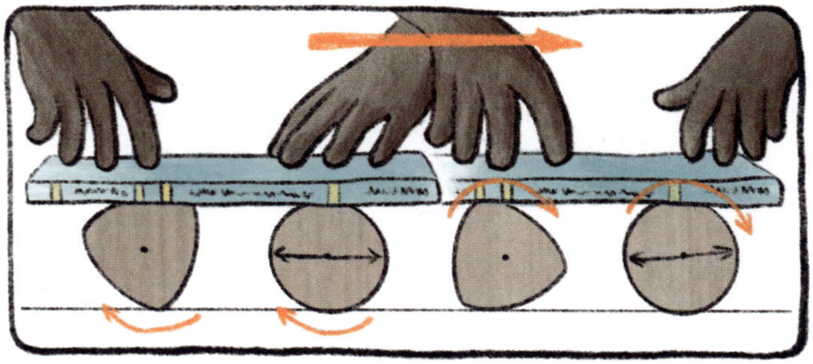

Triangle car

Can you use your wheels to make a toy car?

Curves from nowhere

Can you make a perfect curve using only straight lines? Yes! You can do it right here.

Get your ruler ready ...

All you need is a ruler and a pencil or pen.

Look at the frame below, and you'll see that it's marked 1–20 down the side, and 1–20 along the bottom.

All you have to do is draw a straight line connecting the two dots marked 1… (We've done that one for you!)

Then another connecting the two dots marked 2 …

Then the two dots marked 3 …

And so on!

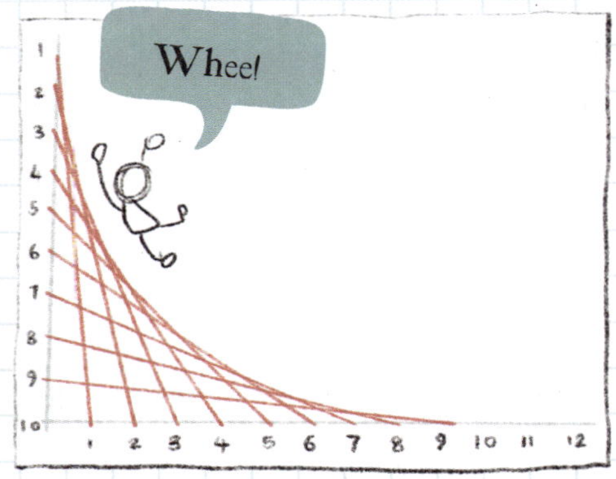

Try these!

Once you've got the hang of it, you can use this method to make more amazing mathematical artworks.

Try a cross with a curve in each corner, like this:

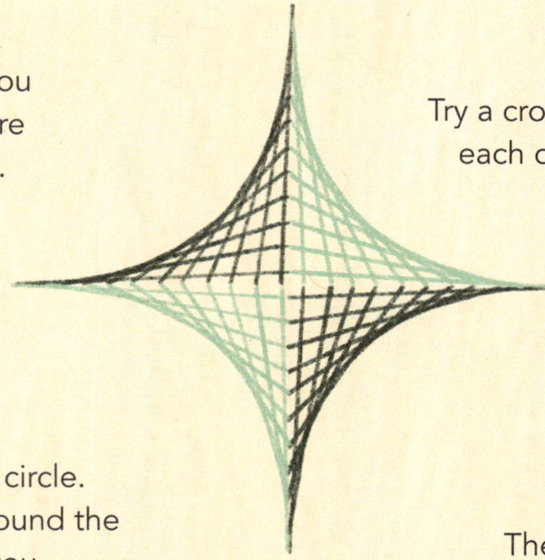

What about a circle?

Draw around something to create a circle. Then mark an even number of dots around the edge using a protractor to help you.

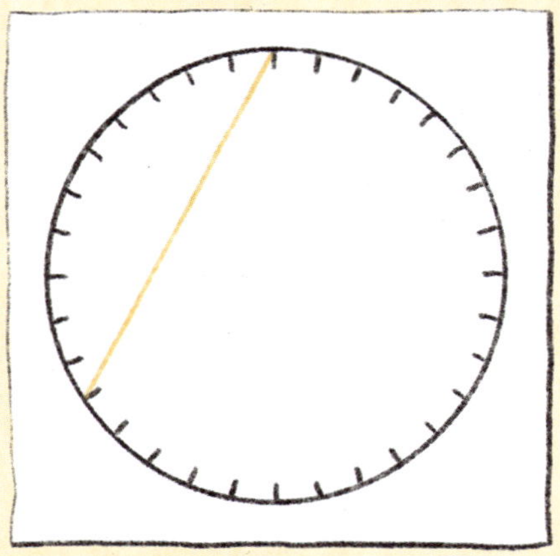

This one has 20 dots

Then connect the dots with lines.

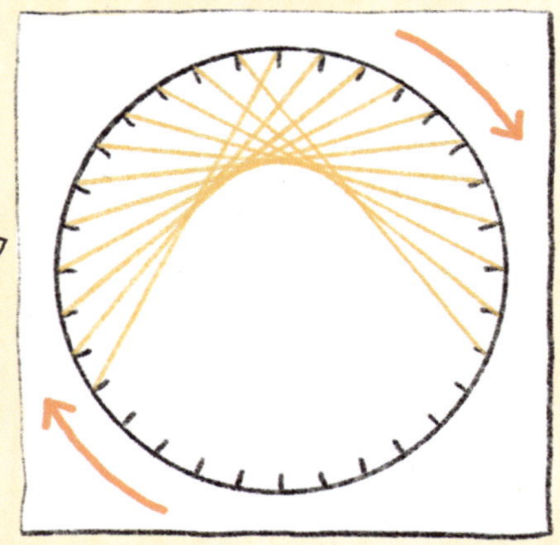

Each dot is connected to the one 8 dots along

You could use rainbow lines to make a pattern

Or fill in the shapes with different shades

Curves of pursuit

Here's another cool pattern you can make with a simple mathematical trick.

It starts with a square ...

First, use a ruler to measure and draw a neat square on a piece of paper. Around 10 cm x 10 cm (4 inches x 4 inches) is a good size to start with.

Next, start in a one corner, measure 1 cm (or ½ an inch) along the line, and make a mark like this.

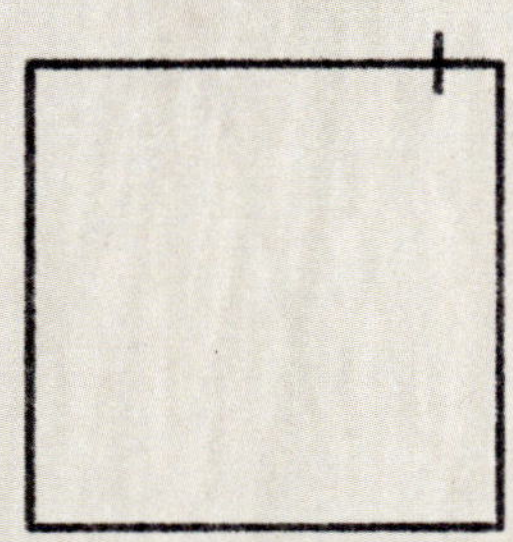

Do this for all the corners, always measuring in a clockwise direction around the square.

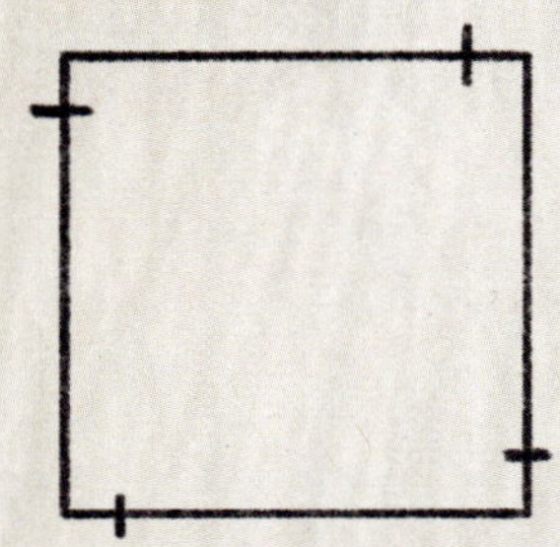

Now draw a new square by connecting the four marks with straight lines.

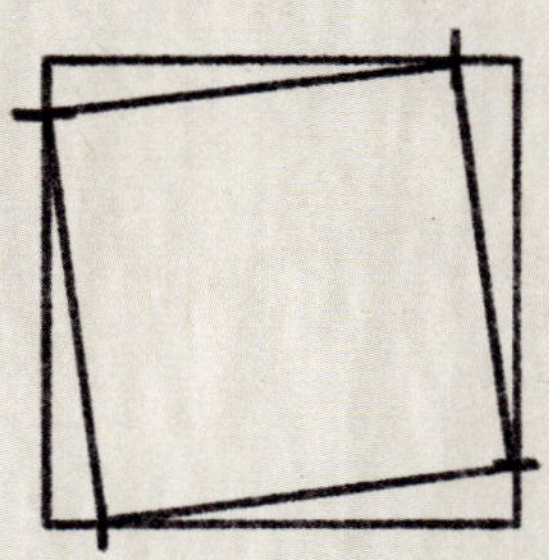

Next, do the exact same thing again, but this time on the new, smaller square …

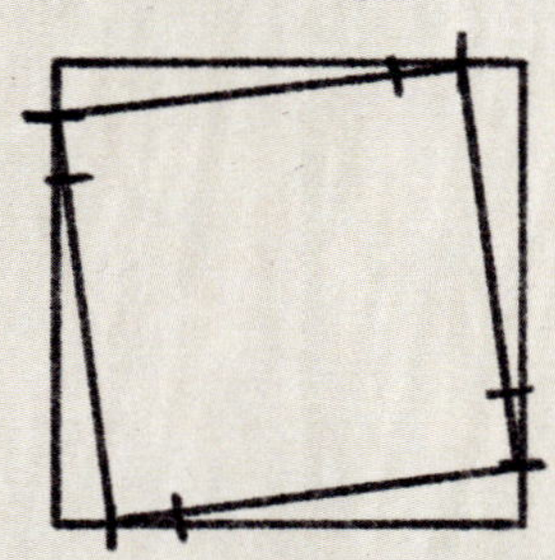

… and connect the new marks to make a THIRD square, like this:

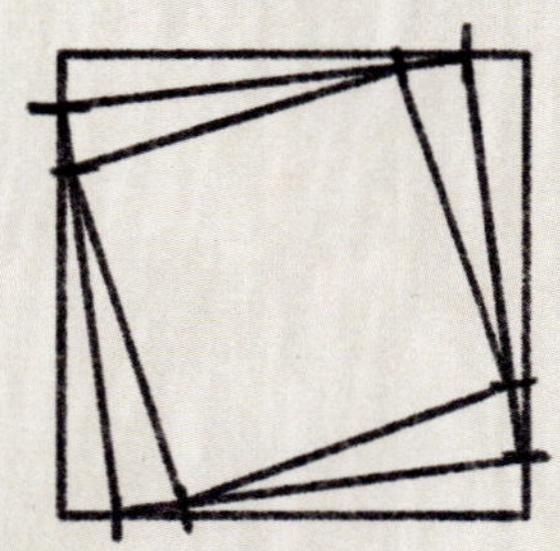

Keep doing the same thing until you can't fit any more squares in, and it will look like this!

It's called a "curve of pursuit."

More shapes

It works with other shapes, too.

Here are some for you to fill in on the page—then try your own designs.

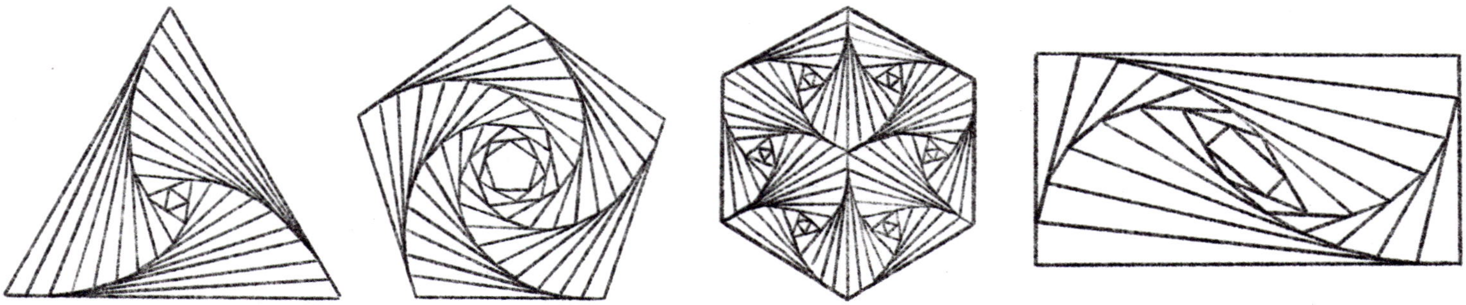

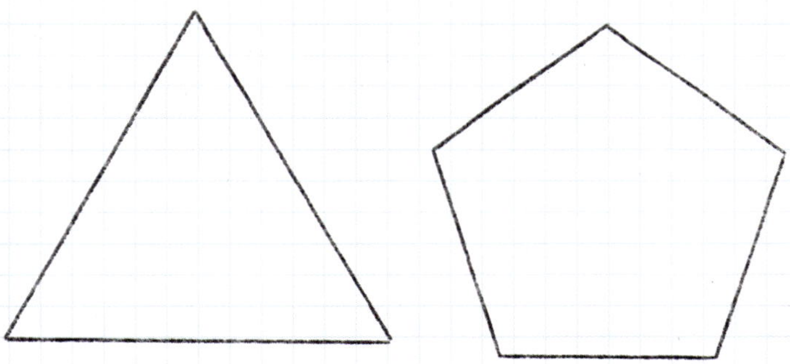

Where's the middle?

It's easy to find the middle of a circle or square.

But where's the exact middle point of THIS shape?

Finding the middle

There is actually a clever way to find the middle of ANY shape. You'll need some stiff card (like an old packing box), scissors, a pencil, a ruler, and a sewing needle or pin.

First, draw a random shape or object on the card, and cut it out.

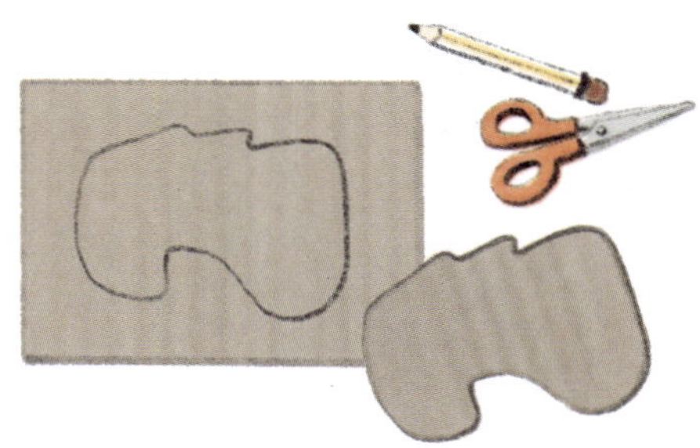

Now push the needle or pin through the shape as close to the edge as you can.

Wiggle it around, so that the needle or pin fits loosely in the hole, and the shape hangs from it freely.

Now carefully hold the shape against a wall in the same position, and draw a vertical line down from the pin.

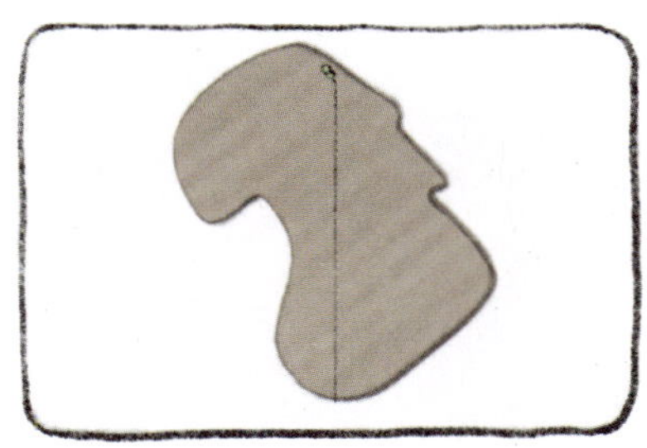

Do the same thing again, making a new hole in a different place on the edge.

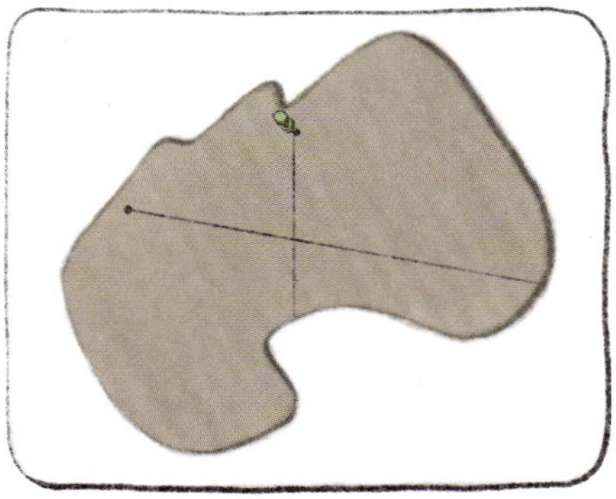

The point where the lines cross is the middle!

Prove it!

To test it out, balance the shape on the middle point, on your fingertip

Shape challenge

Mark where you think the middle points of these shapes are ...

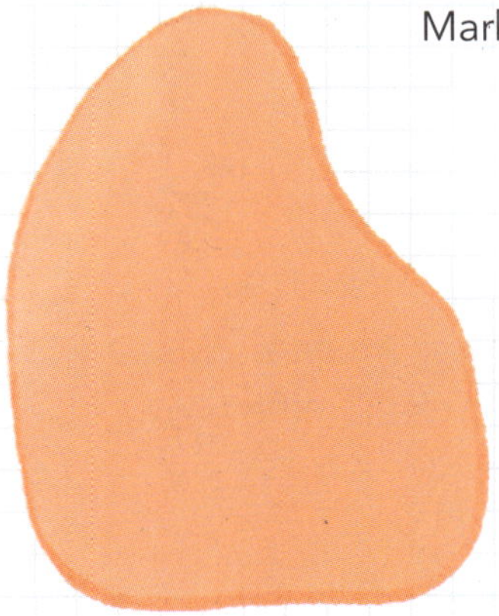

Loop the loop!

Get loopy with some stringy challenges!

Untangle yourself!

Here's an awesome string trick to try with a friend.

Cut two pieces of string, each about 1 m (3 feet) long.

Ask an adult to tie the ends of one string loosely around one person's wrists.

Then do the same for the other person, looping the strings together, like this

Now untangle yourselves without taking off the strings!

Here's how!

Take the middle of the other person's string.

Point it away from you, and push it through the loop around one of your wrists, like this.

Loop it over your hand ...

And you're free!

A knot, or not?

There's a whole branch of mathematics called **knot theory**—all about strings and knots!

This loop has a knot in it

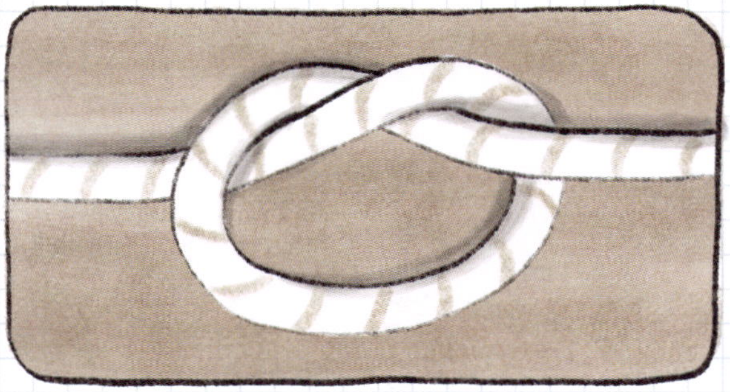

This loop doesn't. If you pulled the sides, it would untangle

Try these!

If you pulled the ends of these ropes, would they have a knot—or NOT?

Answers on page 94!

Dice magic!

For this page, you need at least three dice.

Magic 7s

A dice has six surfaces numbered with 1 to 6 dots.

Look closely, and you'll see that the opposite sides of a dice always add up to 7.

1 is opposite 6

2 is opposite 5

3 is opposite 4

Time for a trick!

You can use this to perform a **mathe-magical trick** on your friends or family.

Ask them to stack three dice on a table, while you look away.

Tell them you can magically see all the hidden sides, and will add them up!

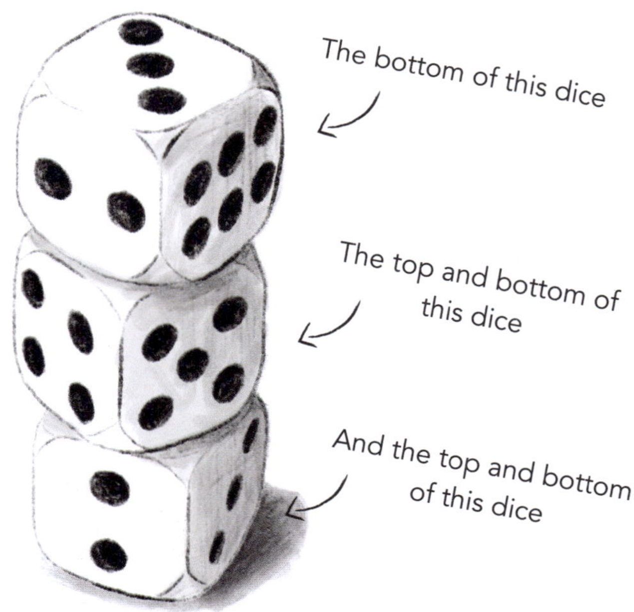

How?

Remember, opposite sides add up to 7.

So the tops and bottoms of all three dice must add up to 3 x 7, or 21.

Subtract the number on top of the stack from 21.

In this example, it's 3

21–3 = 18

Dice number challenge

Now it's the other way around!

We've given you the total of the hidden surfaces—so what should be on top?

Fill in the right number of dots on top of each stack.

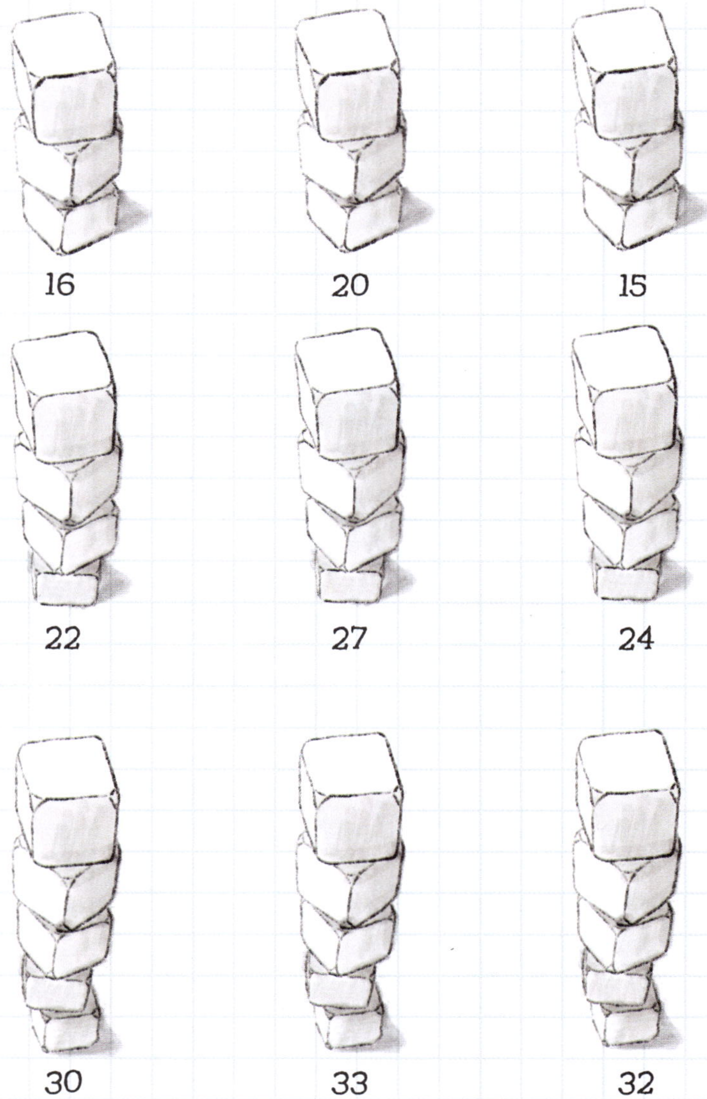

See page 94 for the answers.

Tessellating tiles

If a shape tessellates, that means you can fit lots of that shape together with no gaps, just like tiles covering a floor or wall.

Like this!

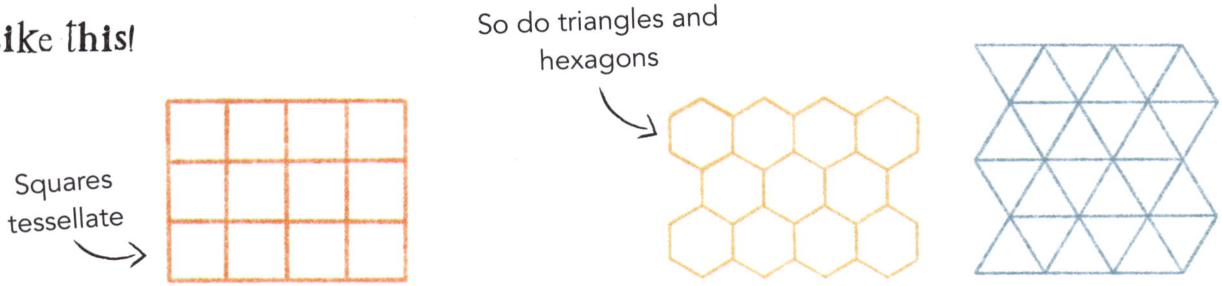

These shapes are very simple, but some much more complicated shapes can tessellate, too.

Make your own!

It's easy to design interesting tessellating shapes.

You'll need some thin card, such as an old cereal box.

Use a ruler to draw a square about 5 cm (2 inches) wide, and cut it out.

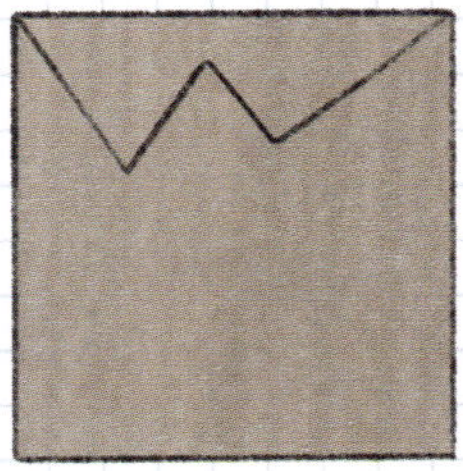

Next, draw a line across it to make an interesting shape.

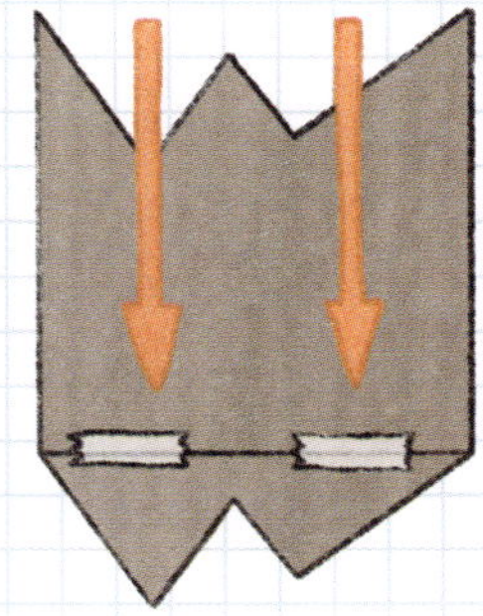

Cut along the line, then swap the two pieces around so the straight top and bottom edges line up. Tape them together, like this.

Take it further by doing the same thing from side to side.

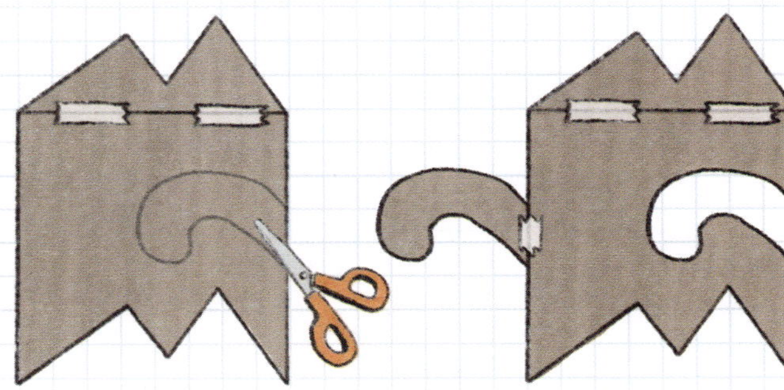

You can now use your shape to draw around onto paper to make a tiled pattern, like this.

Matching Sides

What do you call a shape that's the same on both sides? It's symmetrical!

Like what?

It's like a square, a circle, or a heart. They all have symmetry, meaning that the two sides match.

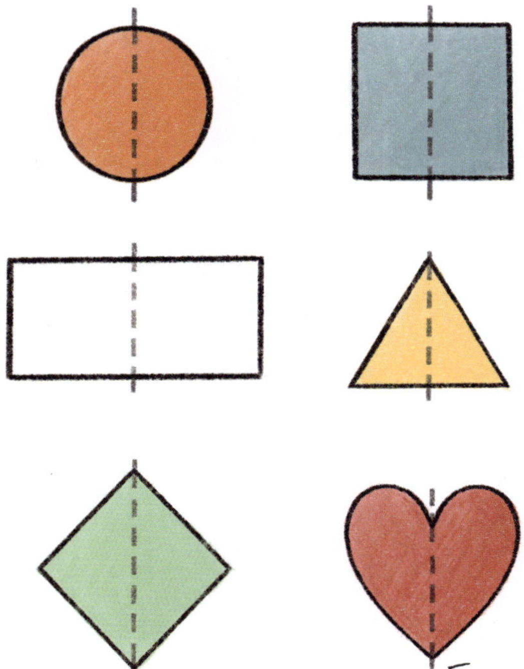

They're like a mirror image of each other

Symmetry everywhere

Lots of everyday objects are symmetrical. Look around your home or classroom, and you'll see some!

T-shirt

Water bottle

Chair

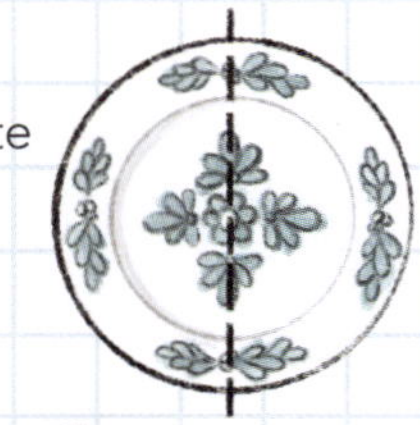

Plate

Butterfly

And you'll find others that are NOT symmetrical, whichever way you look at them …

Sewing machine

Shoe

Calculator

Symmetry self-portrait

To do this, you need a clear photo of your face printed out at a large size.

Carefully cut it in half right up the middle, and stick it onto one half of a blank piece of paper, like this.

Then use pencils or paints to fill in the missing half by copying the photo as accurately as you can.

Fill it in!

Can you draw the missing sides of these symmetrical shapes, so that they match exactly?

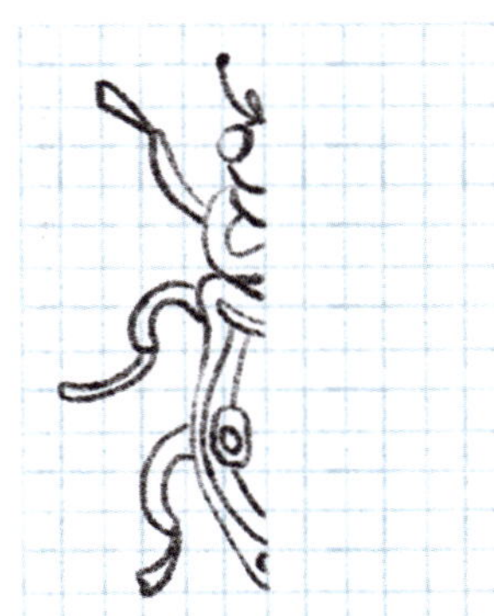

Fract-ivities!

Explore the mysterious world of fractals, with just pencils and paper.

What is a fractal?

A fractal is a mathematical pattern that repeats itself.

To draw it, you just keep adding more parts, following the same simple rule each time.

Here's an example—a stick with two branches.

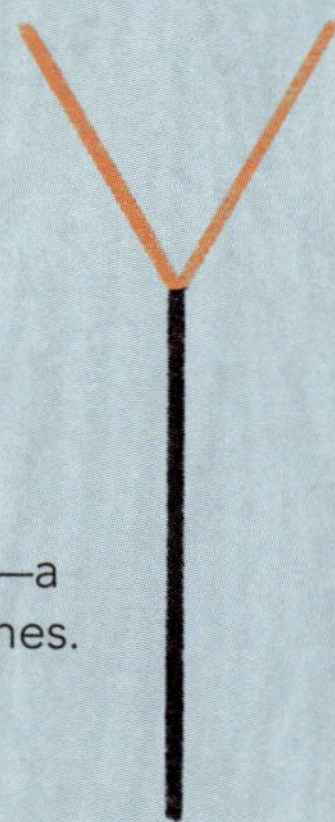

To add to the pattern, you add two more branches to each smaller stick, like this:

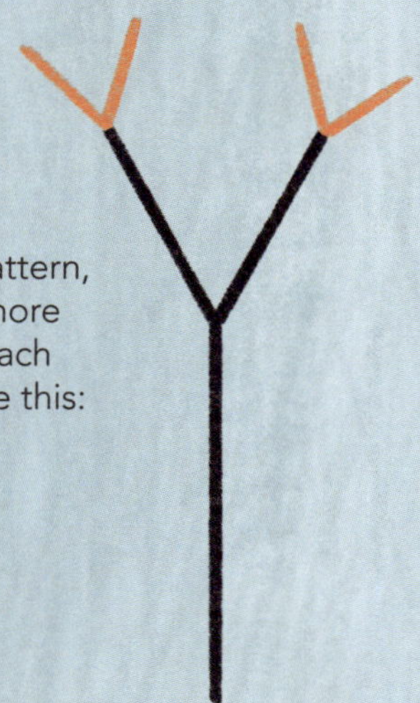

Then keep going …

… and going …

After a while, it starts to look like a tree!

Try these!

Can you add more to these fractals and see what happens?

A fractal H pattern. Add a new H to the end of each stick

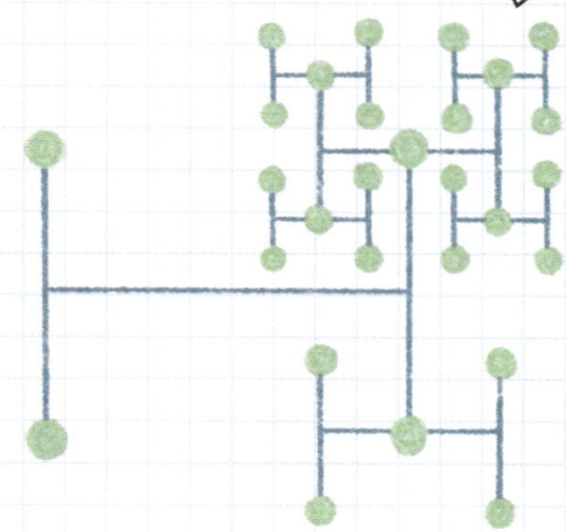

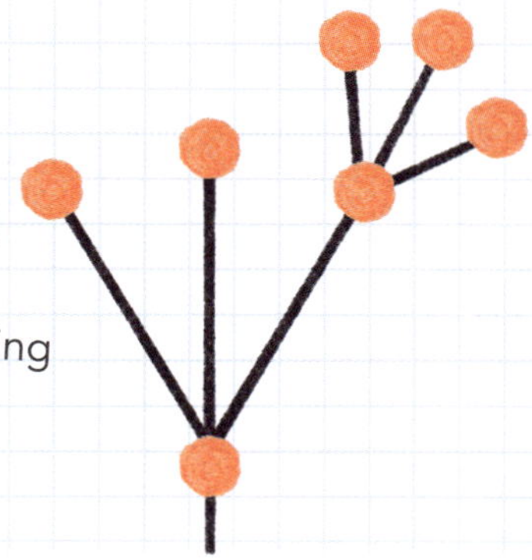

A triple-branching berry plant

Triangle fractal

Fractals don't just grow out—they can grow INSIDE, too!

This one is called a Sierpinski triangle.

Wherever there's a triangle like this

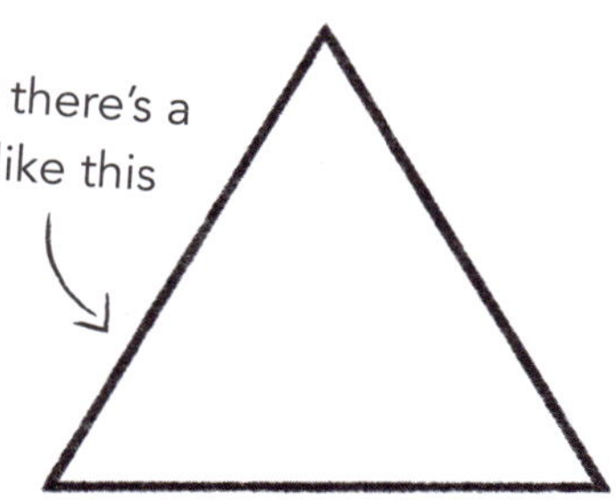

You draw a smaller, upside-down triangle inside

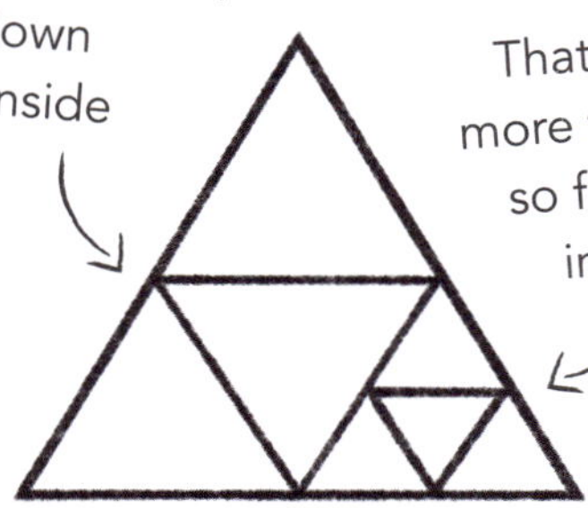

That makes more triangles, so fill them in, too

Square it!

How do you square a number?
You just multiply it by itself.

For example ...

3 squared means 3 x 3. That makes 9, which we call a "square number."

3 x 3 means 3 3s ...
which looks like this

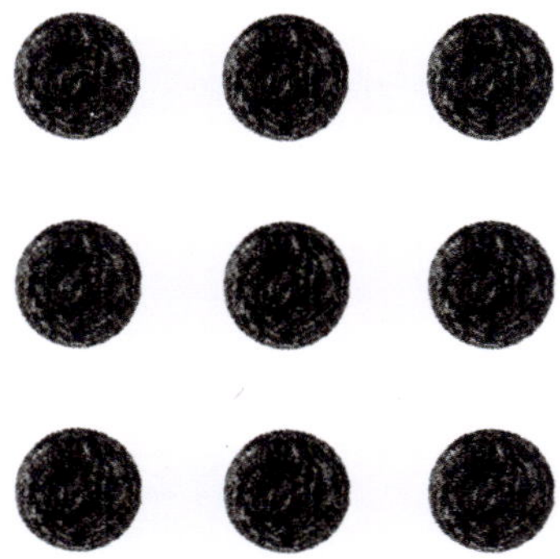

1 X 3

2 X 3

3 X 3

And now you can see why it's called a square number!

The dots form a square.

Here are some more square numbers ...

2 X 2 = 4

4 X 4 = 16

5 X 5 = 25

10 X 10 = 100

The squares game

Now that you know what a square number is, you can play the square numbers game.

You need a pencil, a ruler, and graph (squared) paper like this.

Draw a rectangle on the grid. It can be any number of squares tall and wide—as long as it's NOT a square!

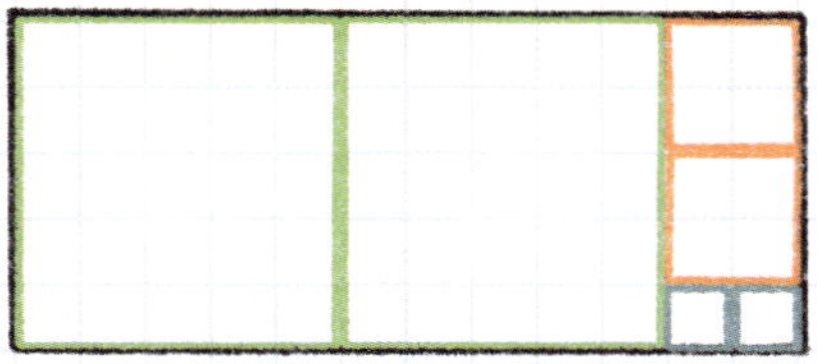

Now you have to fill in the rectangle with squares, using as FEW as you possibly can—like this! We did it in 6 squares.

Here are some more to try …

And some space for doing your own

Two players

To play this with a friend, take turns filling in the squares. Whoever draws the last square wins!

Drawing in 3D

How do you draw a cube, box, or house so that it looks real and 3D?
Mathematical magic!

The simple version

First, try this.

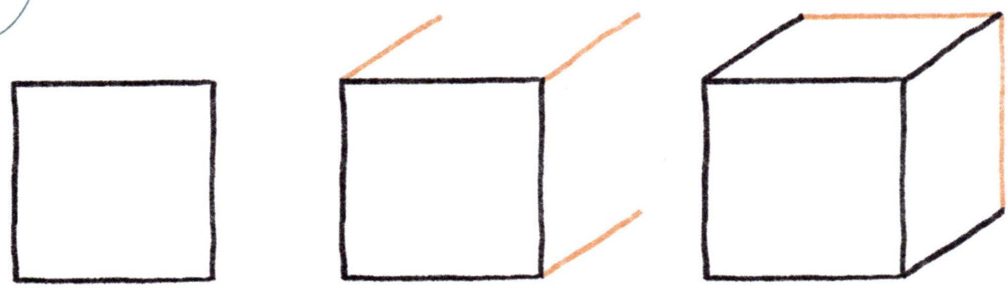

Draw a square.

Add three diagonal lines, all the same length, to these corners, like this.

Now join up the lines like this.

There you go!

It works with a rectangle, too ...

You can use them to draw all kinds of things.

Into the distance

Now try this method for an even more amazing effect!

Draw a straight line across a piece of paper, and add a dot.

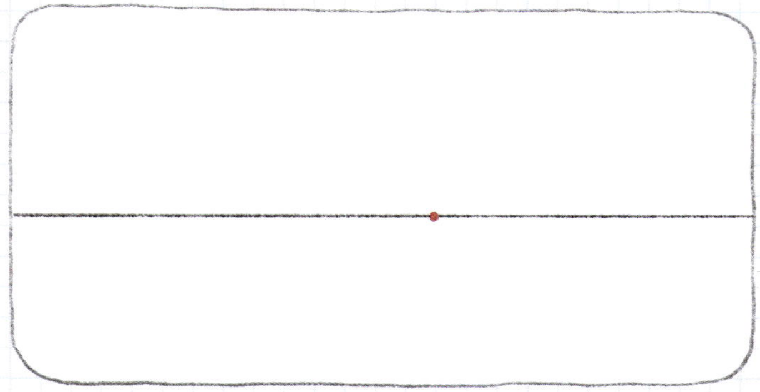

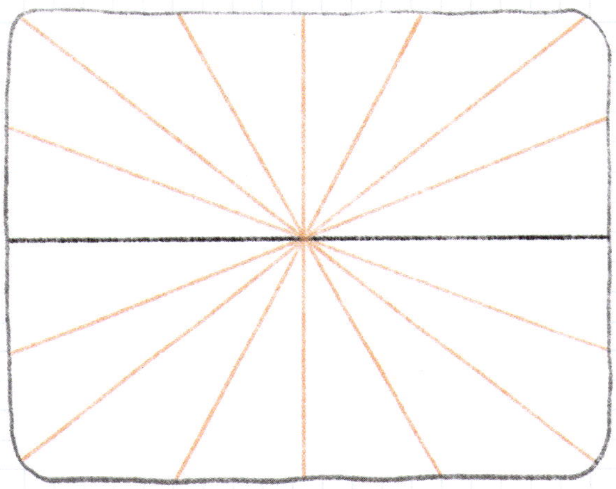

Draw faint lines from the dot to the edge of the paper.

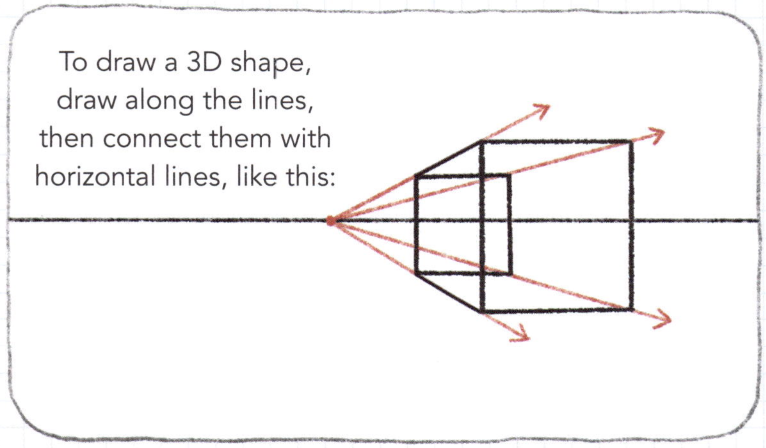

Just add more lines wherever you need them … Then erase them when you're done.

You can draw a whole scene this way!

Easy as Pi

Have you heard of Pi? Not pie, Pi!

What is Pi?

Pi is simply the circumference of a circle …

… divided by its diameter.

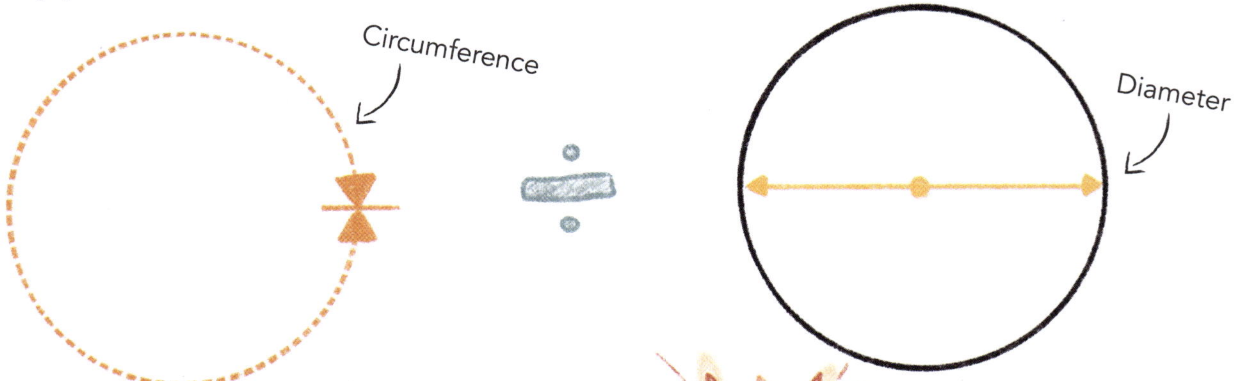

Whatever size your circle is, the answer will always be around **3.14** or roughly three and a seventh. And this number is called Pi!

See for yourself!

To test this, find a round object, such as a roll of tape …

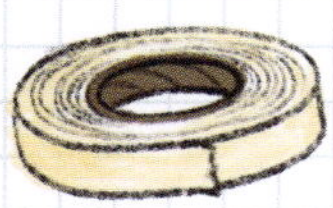

… and some string and scissors.

Wrap string around the object to measure its circumference, and cut it off.

Now use this string to measure the diameter across the object.

Snip it into diameter-sized pieces.

How many are there? About 3 and a 7th, right?

This will always happen—with ANY circle!

Pi numbers

Pi is a decimal number somewhere in between 3 and 4. In fact, it goes on forever!

3.14159265358979323846264338327950288419716939937510 ... And so on!

Pi pictures

You can use Pi to make cool pictures.

For a Pi city picture, use the numbers to decide how tall each skyscraper is.

This is 3 cm (or inches) high, and so on!

Pi people

Can you use Pi to draw a picture of a row of people or animals of different heights?

3.   1   4   1   5   9   2   6   5   3

Measuring wheel

You can use a circle to measure distances with a measuring wheel.

Make a mini measurer

To make a measuring wheel, you need thick card and a split pin or brass fastener, like these.

Draw around a large coin onto the card, and cut the circle out.

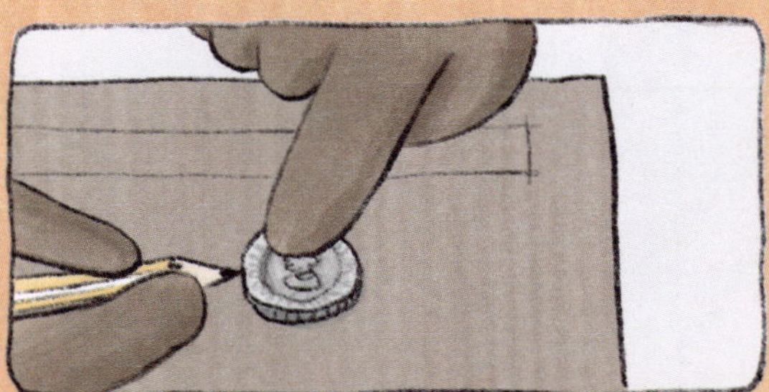

Draw a dot in the middle of your wheel, and another near the end of the handle.

Push a fastener through both dots to hold them together.

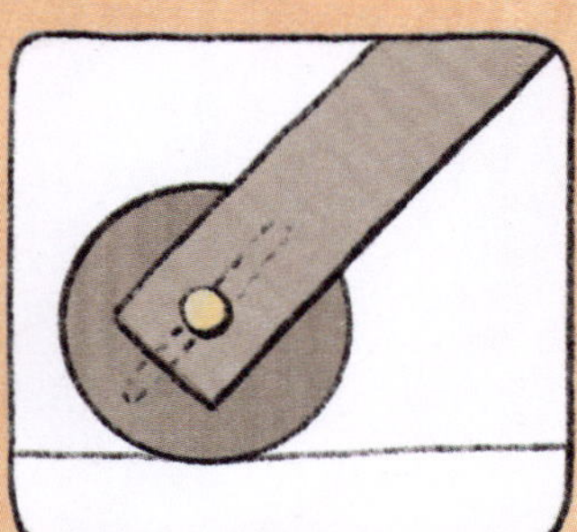

Draw a red dot on the edge of the wheel.

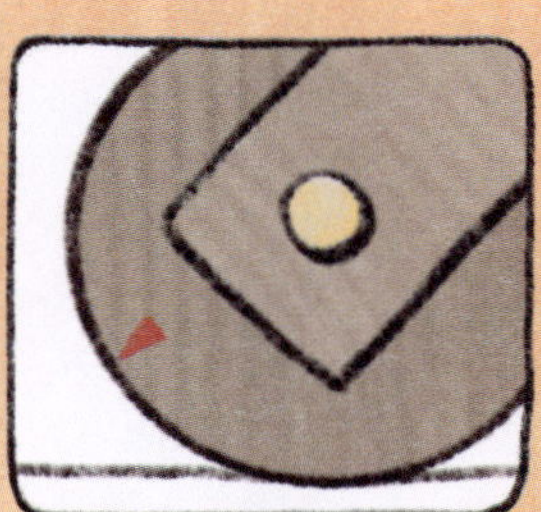

Now roll the wheel along on a piece of paper, and mark the distance of one full turn.

To measure a distance, roll the wheel along, and count how many times it turns.

Then multiply this number by the length of one turn.

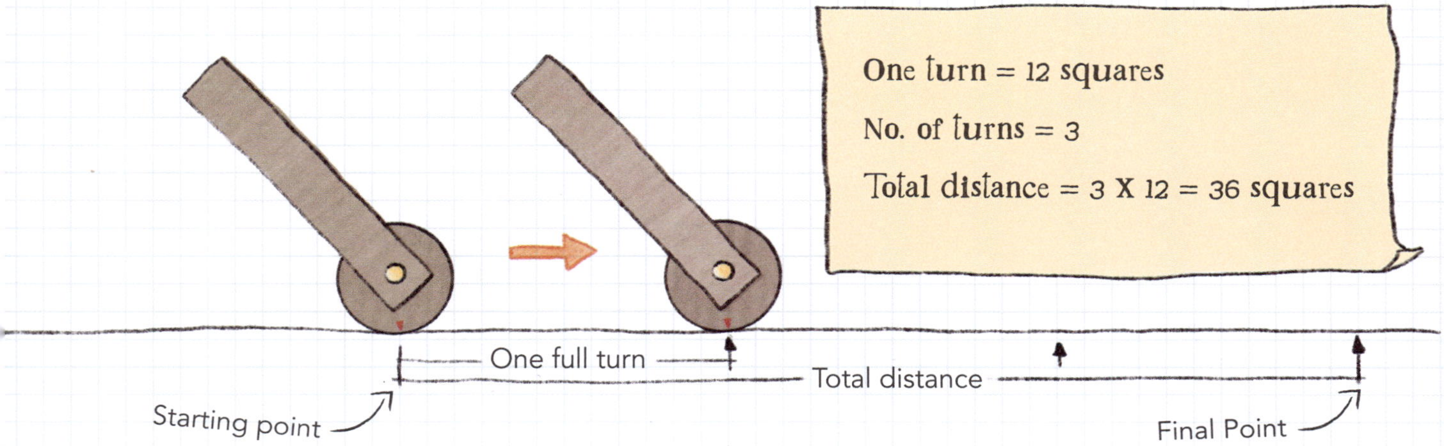

Try out your wheel on our map.

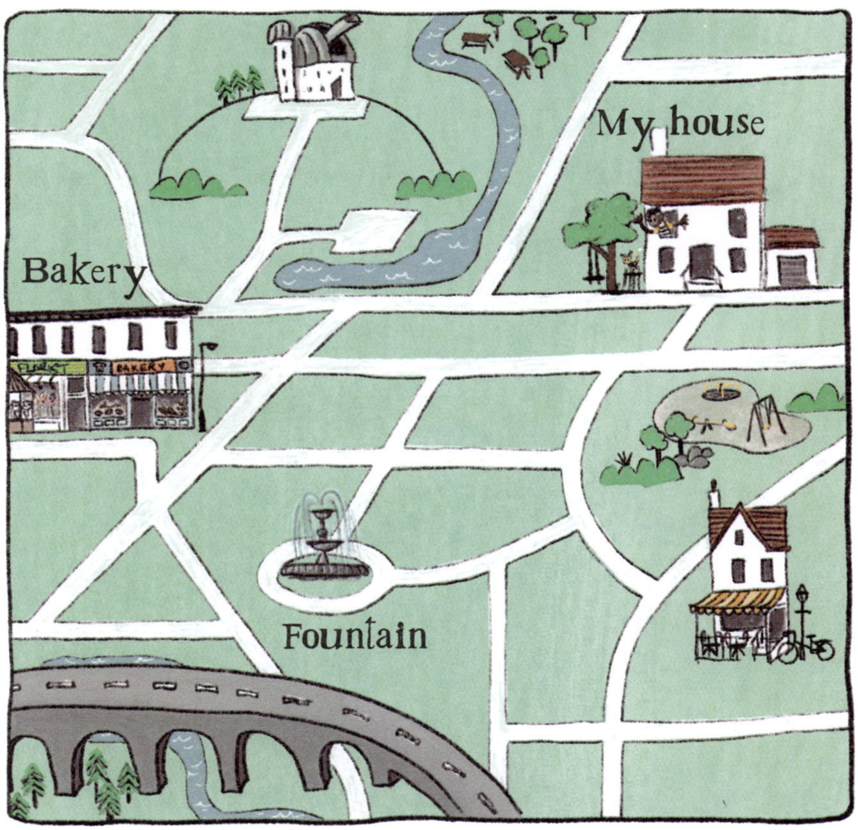

X marks the spot!

You're trying to read a map, but you can't find what you're looking for.

You need coordinates!

Find that street!

Here's a map. Quick as you can, find Egg St. Where is it?

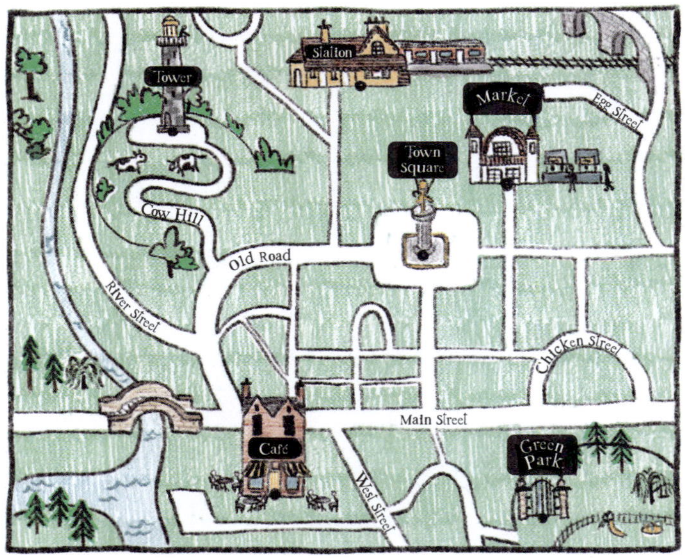

Unless you were lucky, it probably took a while, because you have to look all over the map!

Squares on a map

Coordinates make it easier. They divide the map into squares.

Each row and column of squares has a letter or number

So by combining these, you can give each square a name.

This one is A1

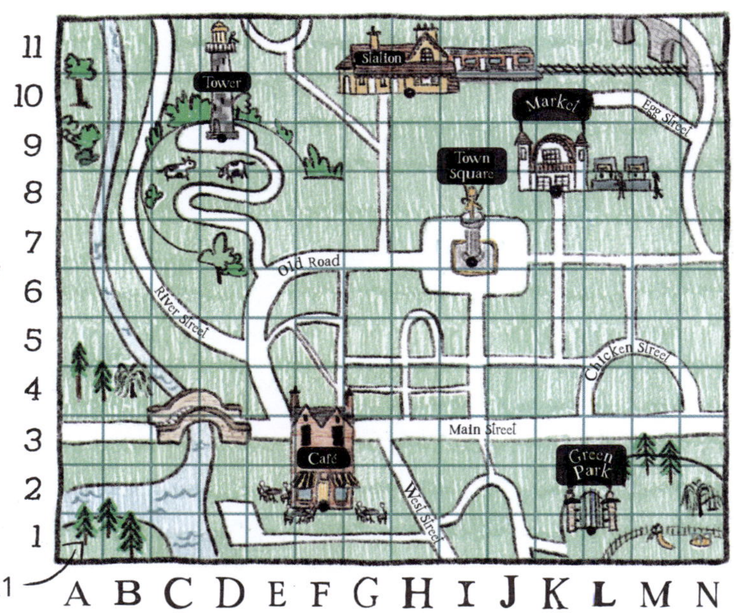

Let's try it again.

Can you find the tower?

It's in square D9.

Mapmaker

Here's a grid to draw on.

Draw a network of streets, and add buildings, landmarks, or other places.

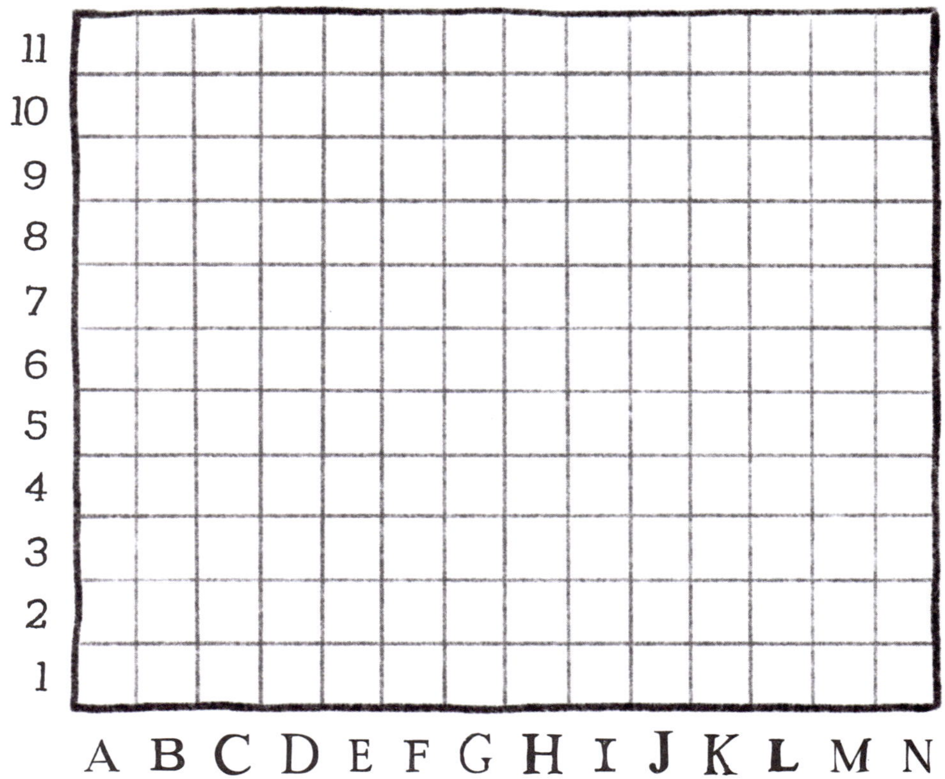

Can you put ...

- A street named Letter St in square E9
- A statue in square D2
- A library in square D6
- A playground in square C7
- A duckpond in square B4

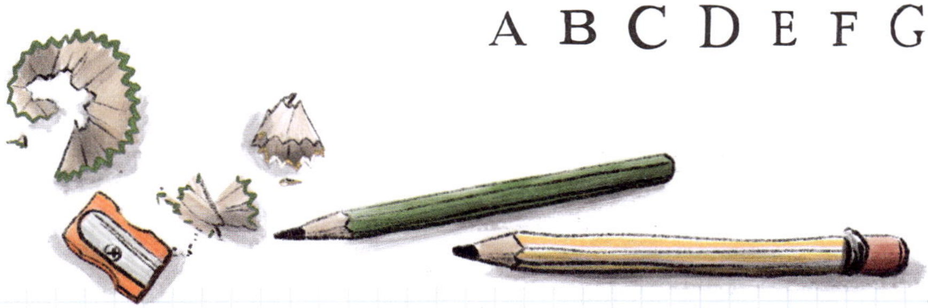

Add your own features and locations, too! You can list them here:

____________________ Square: ________

____________________ Square: ________

____________________ Square: ________

____________________ Square: ________

Pixel art

Pixels are little squares that can be used to make up a picture.

Computer screens use tiny pixels in their displays

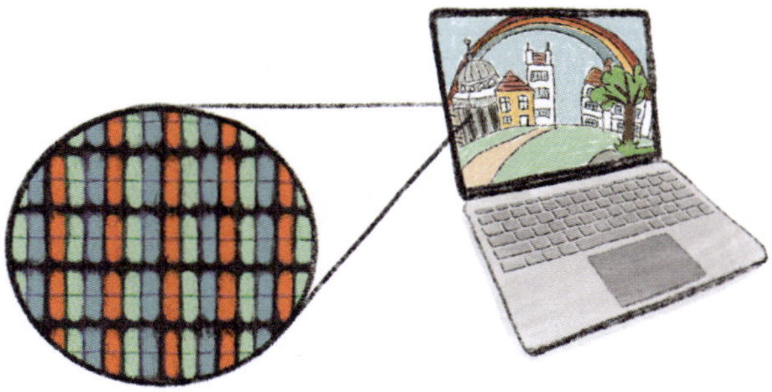

Drawing in pixels

To make a picture look like it's made of pixels, you need some graph (squared) paper.

You can buy it, or print some out from the Internet.

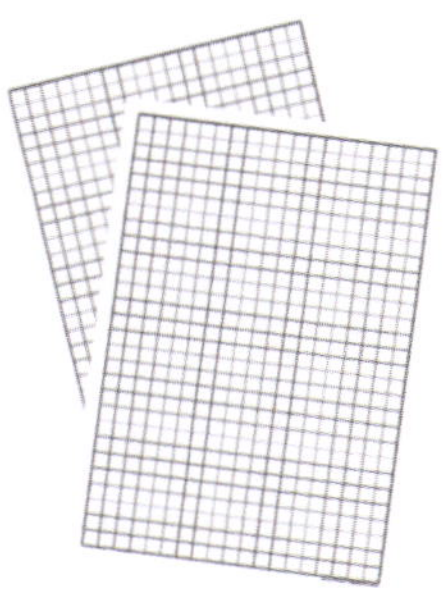

Or draw a grid using a pencil and ruler

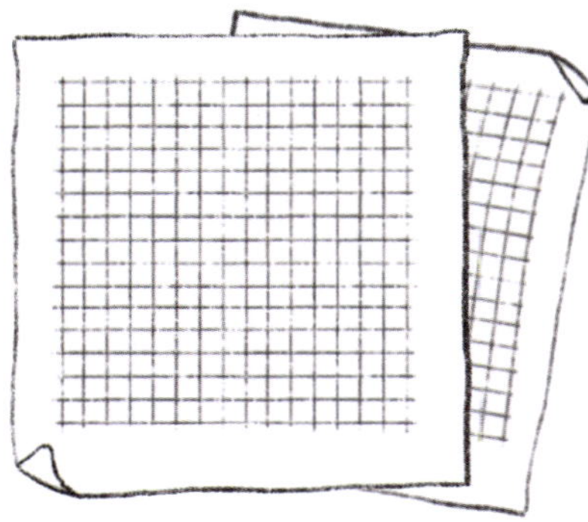

Small squares work best!

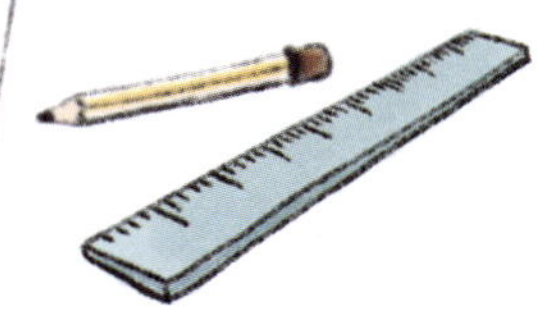

If you like, just start drawing straight onto the squares.

Use one pen or pencil to fill in each whole square

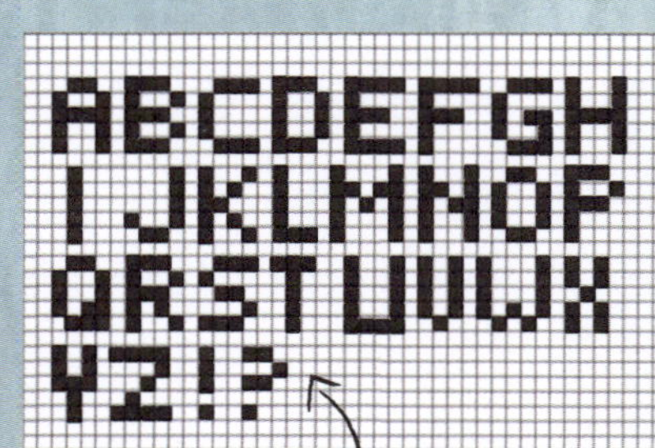

And keep filling them in to make patterns or pictures

You can do pixel letters, too

Use a sketch

If you're not sure where to put your pixels, try drawing a sketch first in light pencil.

Use it as a guide, then erase any lines you don't want.

Try it!

To get started, try copying this pixel picture into the grid squares next to it.

Then use this space to try out your own ideas!

What next?

Can you look at a string of numbers and see what comes next?

Spot the sequence

All you need to do is look at the numbers to see what's happening, then figure out what's next.

This one's pretty easy!

1	2	3	4	5	6	7	__	__	__

You just add 1 each time—so the next numbers are 8, 9, and 10.

Now try this:

2	4	6	8	10	12	__	__	__

In this sequence, you add 2 each time.

Now it's getting harder ...

2	3	5	6	8	9	10	__	__	__

In this one, you add 1, then 2, then 1, then 2, and so on. Did you spot it?

How about this?

0	1	1	2	3	5	8	13	__	__	__

This is a tricky one. Each time, you add the two previous numbers to get the next number.

This is a famous number sequence named the Fibonacci sequence after mathematician Leonardo Fibonacci

Try making up your own number sequences to test your friends and family. Can they crack them?

Sequence spirals

You can use number sequences to make spirals on graph paper.

For example ...

1 2 3 4 5 6 7 8

Take the first number, and draw a line that many squares long.

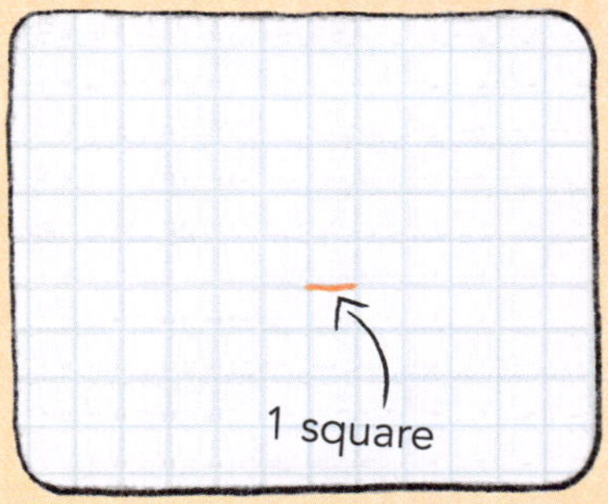

Then turn a corner, and do the same with the second number ...

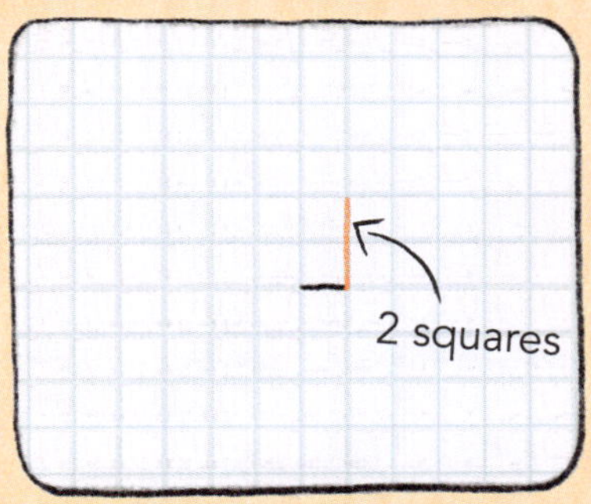

... and keep going!

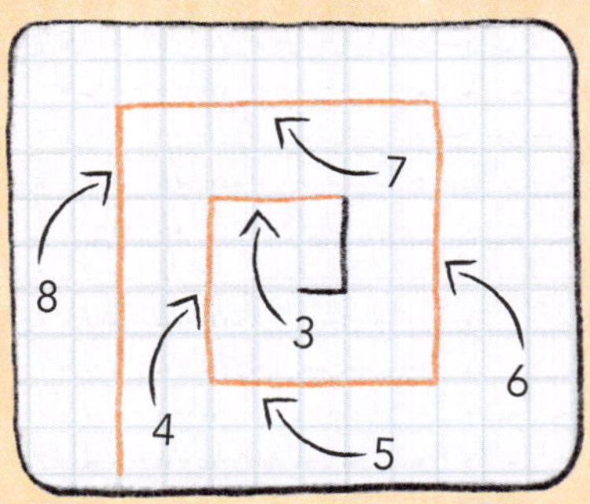

2 3 5 6 8 9 10 12

1 1 2 2 3 3 4 4 5 5

0 1 1 2 3 5 8 13

See page 94 for the answers.

Nets and Shapes

Instead of catching fish, in mathematics, you use a net to make a 3D shape.

How?

Here's a simple one to start with.

How would you make a shape like this out of paper or card?

This is a simple 3D pyramid shape—just like the famous pyramids of Giza in Egypt.

You'd use a net—and here it is!

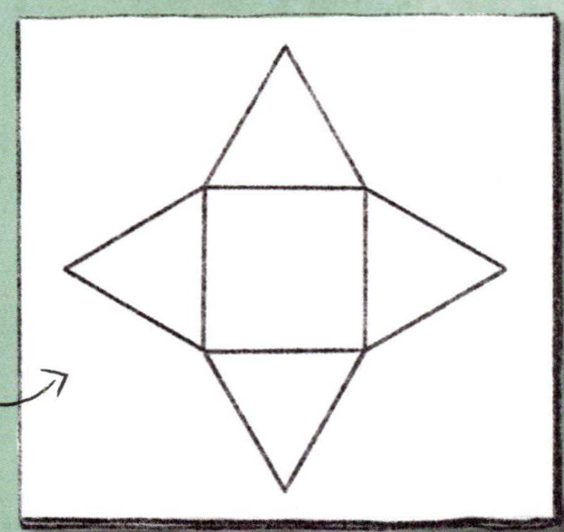

If you cut this net out and folded it along the lines, it would fold up into a pyramid shape.

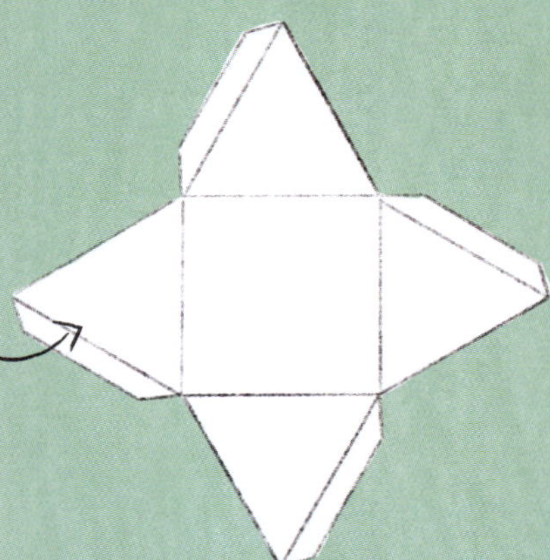

And if you add little tabs at the edges of the net, you can glue them to hold the shape together

More nets

For each 3D shape, there's a net that makes it.

Here are some more ...

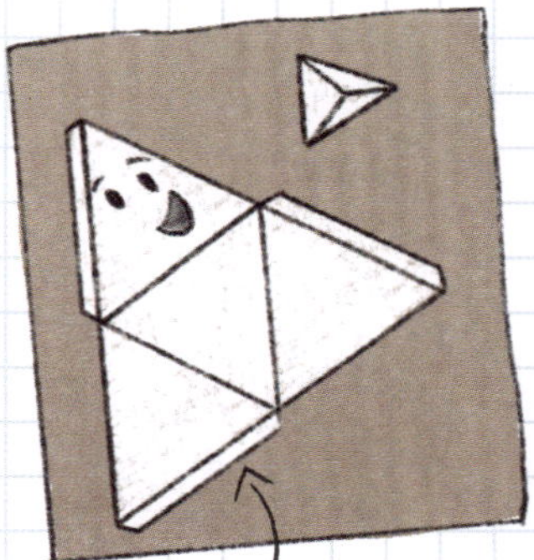

Tetrahedron
(a 3-sided pyramid)

Cuboid

Cylinder

Cube

In fact, there are several different nets for a cube.

Can you think of any others?

In reverse!

You can use a net to make a shape—and you can also look at a shape to figure out what its net would look like.

Cone

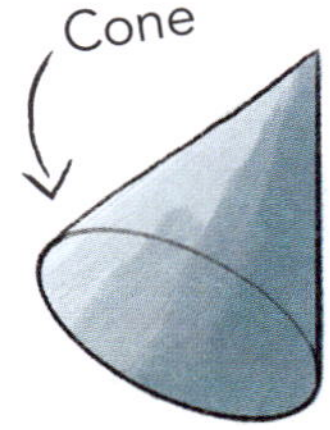

Prism

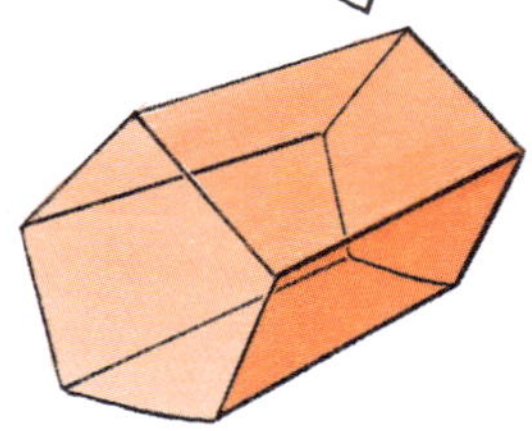

Diamond

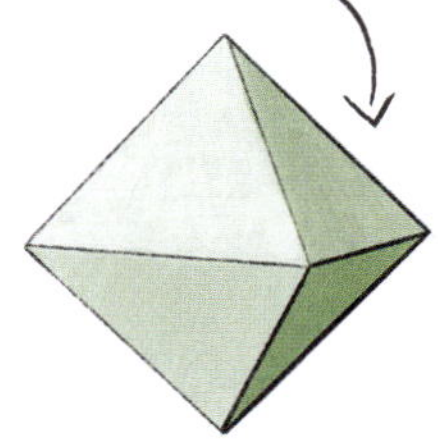

See page 94 for the answers.

Triangle tricks

Triangles are great, and you can do all kinds of clever things with them.

Types of triangles

Are all triangles the same? No! They come in several types.

In an equilateral triangle, all the sides are the same length

In an isosceles triangle, two of the sides are the same

In a scalene triangle, the sides are all different lengths

And this is a right-angled triangle. One of its corners is a right angle, like the corner of a square

Slice it up!

But whatever type of triangle you have, you can always split it into two smaller triangles … like this!

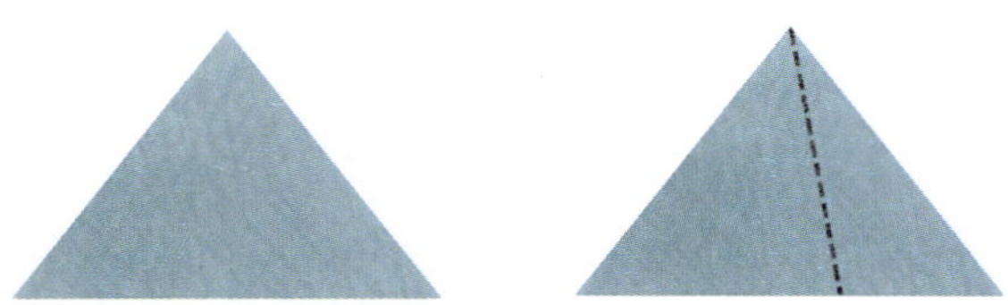

And you can divide them into even smaller triangles …

And so on!

Triple triangle tricks!

Not one, not two, but THREE triangle tricks for you to do!

1 Fill in the triangle with the numbers below it, so that all three sides add up to 8.

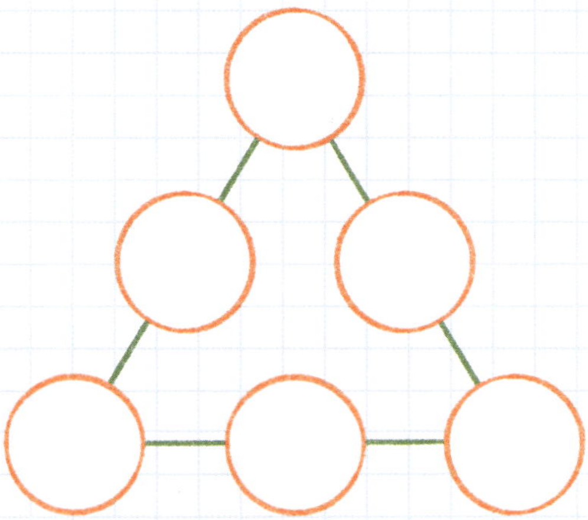

0 1 2 3 4 5

See page 94 for the answer.

2 Add straight lines to make more triangles inside this triangle. How many can you make?

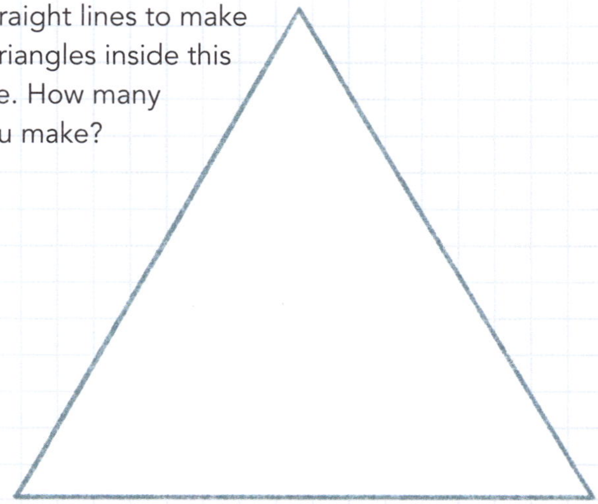

3 How many triangles are there here? (Shading them in might help you count!)

See page 94 for the answer.

Triangular numbers

You've already met square numbers on page 26.

Now it's the triangles' turn!

Building triangles

A square number can be arranged as a square, and a triangular number can be arranged in a triangle.

To see how it works, find yourself some coins, buttons, beads, or other small objects, or cut some small circles out of paper or card.

Make a group of three like this:

You've made a triangle! As you might expect, 3 is a triangular number

So how many do you need to add to make a bigger triangle?

Did you make this?

6 is also a triangular number

And to make the next triangle, you'll need 10.

It's a sequence!

The triangular numbers make a simple number sequence, like the ones on page 38.

Each time, you add one number more than the last time.

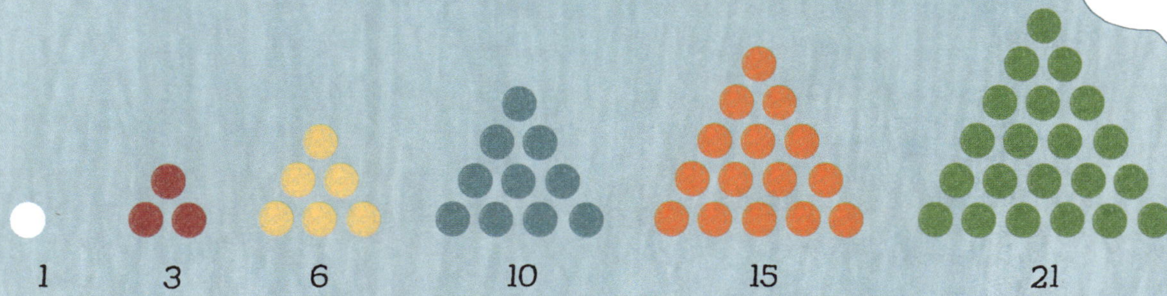

Triangle tree

This triangle grid makes it easy to draw triangular numbers on paper.

Add more triangles until you get to the bottom, and you'll have drawn a pine tree!

What's the triangular number of the completed tree?

Triangles and Squares

Remember what a square number is? Good—you'll need it!

Start with a triangle ...

First, you need a right-angled triangle. Here's one!

In this triangle, the sides are 3, 4, and 5 sqares long.

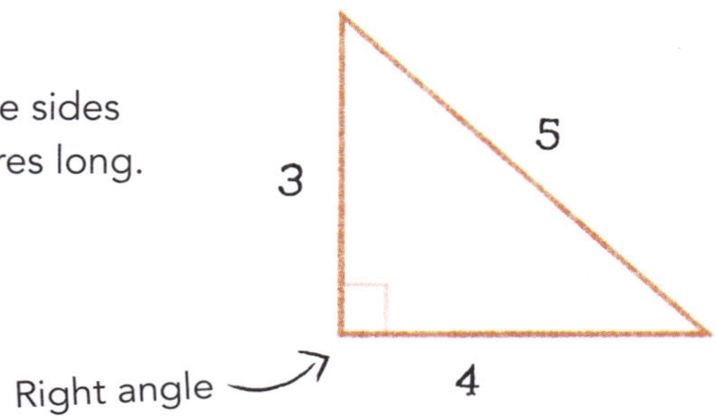

In ancient times, people noticed something strange about triangles like this.

It's called the "Pythagoras theorem" after ancient Greek genius Pythagoras, but other people spotted it, too.

So what is it?

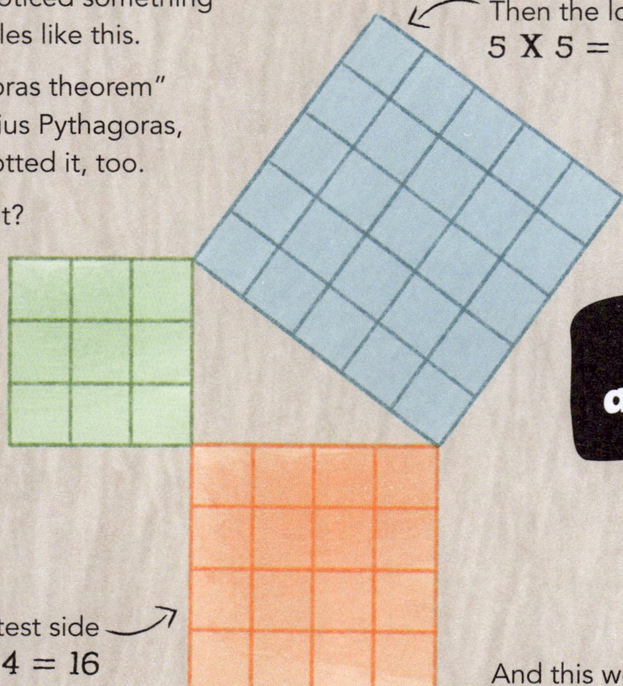

First, square the shortest side
3 X 3 = 9

Then the next shortest side
4 X 4 = 16

Then the longest side
5 X 5 = 25

The first two square numbers add up to the last one!

9 + 16 = 25

And this works for *any* right-angled triangle.

You can use the Pythagoras theorem to solve problems such as ...

Farmer Fallow's fence

Farmer Fallow wants to divide this field in two to keep his ducks and chickens apart.

How long will the fence be? You can think of the fence in yards if that's easier for you.

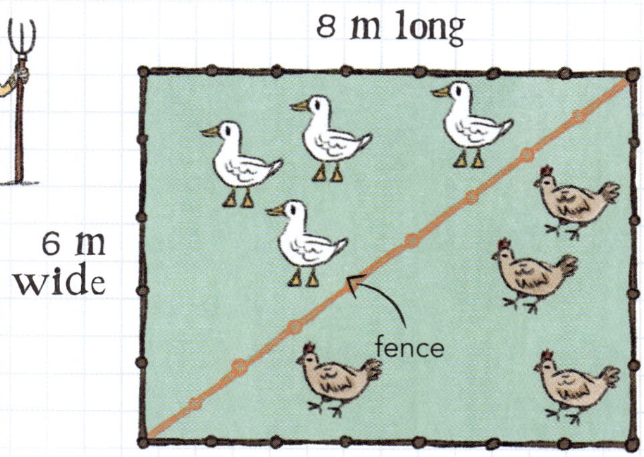

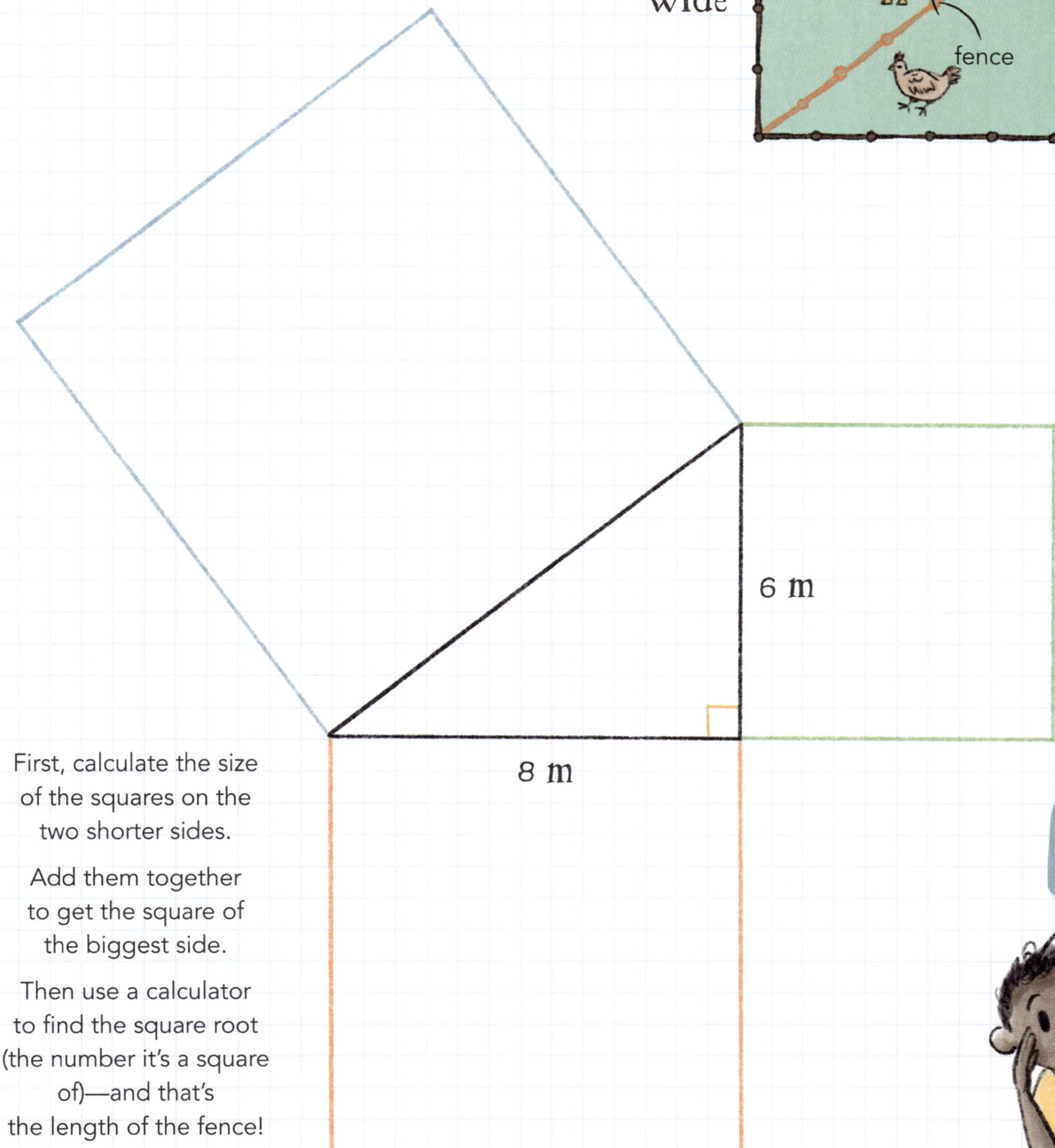

First, calculate the size of the squares on the two shorter sides.

Add them together to get the square of the biggest side.

Then use a calculator to find the square root (the number it's a square of)—and that's the length of the fence!

Hexagons

A hexagon is a shape with six sides, and hexagons can be very useful. Just ask a bee!

Draw a hexagon with a compass

If you have a compass for drawing circles, you can draw a hexagon!

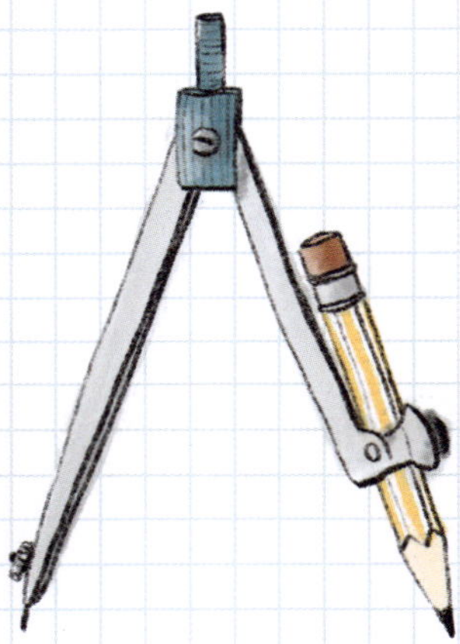

Draw a circle (any size) on a piece of paper.

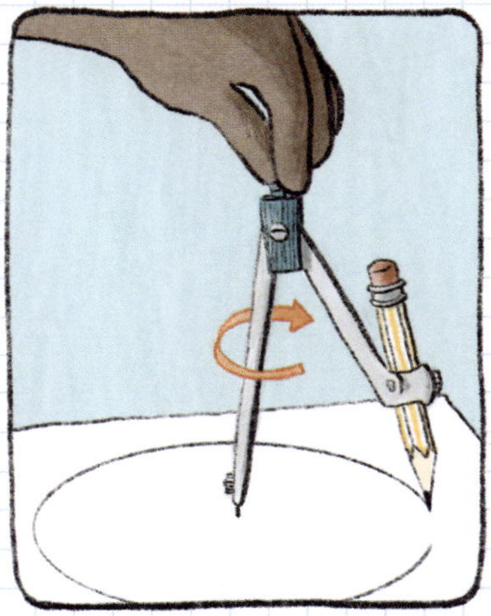

Keeping the compass at the same angle, put the point down on the edge of the circle, and draw a mark farther along the edge where the pencil lands—like this:

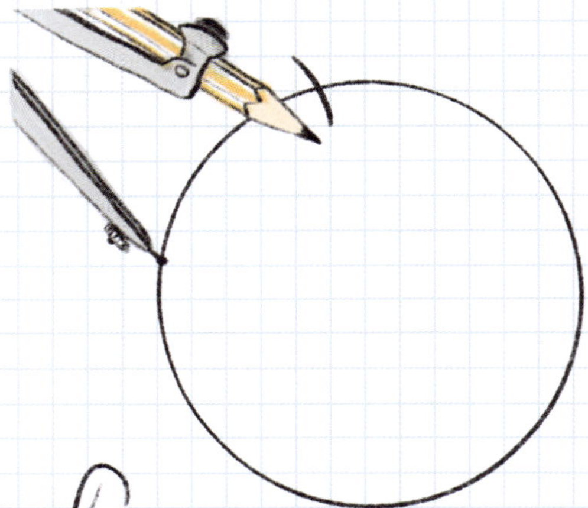

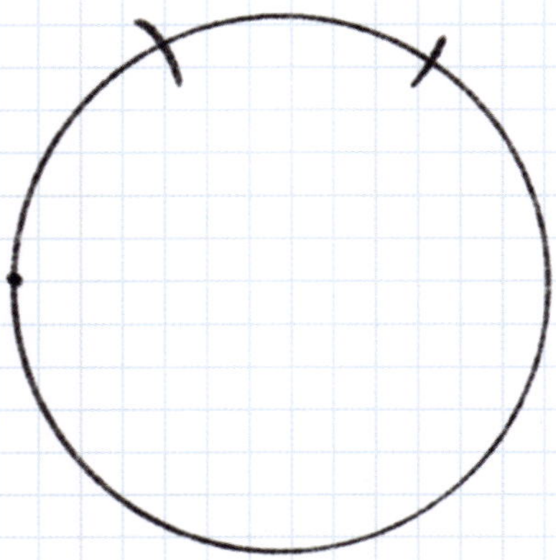

Move the point to the mark you drew, and do the same thing again.

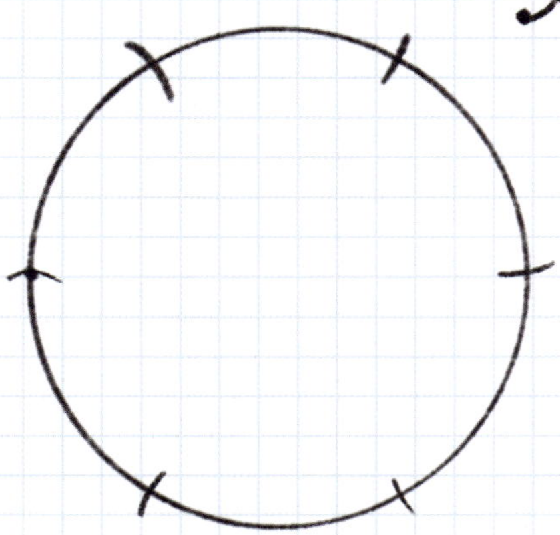

Keep going until you get back to where you started, and have six marks.

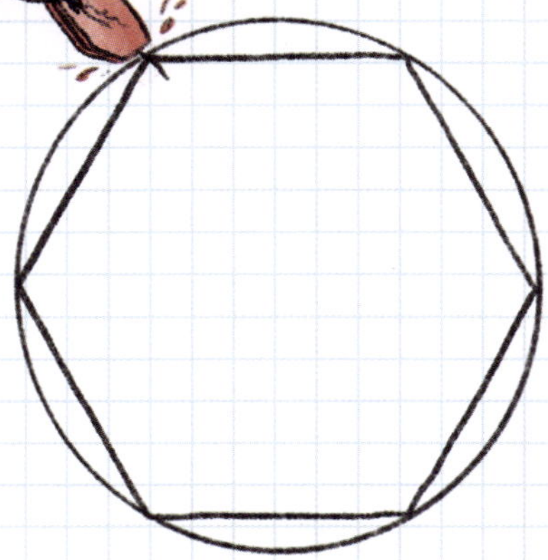

Connect the 6 points with straight lines, and you have a hexagon!

Fitting together

Hexagons tessellate, or fit together like tiles (see page 20).

That's why bees use them when they build honeycomb out of wax.

The spaces, or cells, are used to store honey or other food, and as nests for baby bees

Draw a beehive

Can you make a small hexagon from card, then draw around it lots of times to make a hexagon honeycomb picture here?

Times table table

Times tables can be hard to learn and remember.
You need a times table table!

One two is two ...

Sometimes it seems like there are endless times tables to learn ...

However, you can put them all in one place, making it easier to see them.

It looks like this:

To find any answer, you just find the two numbers you want to multiply, and follow the row and column along until you find the square where they meet—a little bit like the coordinates on page 34.

For example, here's 8 x 9

X	1	2	3	4	5	6	7	8	9	10
1	1	2	3	4	5	6	7	8	9	10
2	2	4	6	8	10	12	14	16	18	20
3	3	6	9	12	15	18	21	24	27	30
4	4	8	12	16	20	24	28	32	36	40
5	5	10	15	20	25	30	35	40	45	50
6	6	12	18	24	30	36	42	48	54	60
7	7	14	21	28	35	42	49	56	63	70
8	8	16	24	32	40	48	56	64	72	80
9	9	18	27	36	45	54	63	72	81	90
10	10	20	30	40	50	60	70	80	90	100

Times table patterns

This isn't just useful—it's also full of cool patterns.

To see them, grab some pens or pencils, and try this:

1. What happens if you fill in all the ODD number squares?

X	1	2	3	4	5	6	7	8	9	10
1										
2										
3										
4										
5										
6										
7										
8										
9										
10										

2. What about if you fill every number square with a ZERO in it?

X	1	2	3	4	5	6	7	8	9	10
1										
2										
3										
4										
5										
6										
7										
8										
9										
10										

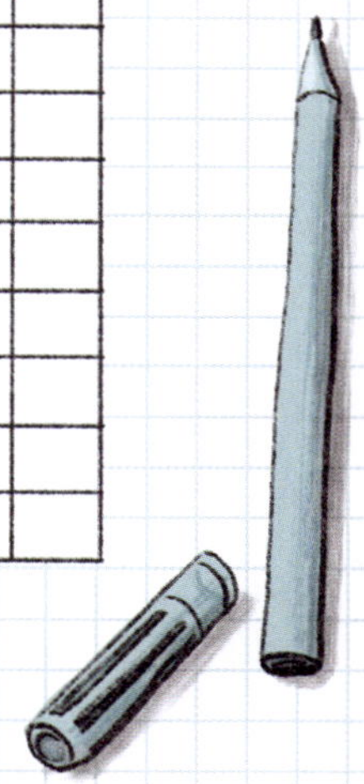

3. Now try filling in all the SQUARE number squares—1 x 1, 2 x 2, 3 x 3, and so on.

X	1	2	3	4	5	6	7	8	9	10
1										
2										
3										
4										
5										
6										
7										
8										
9										
10										

4. Search for your own patterns, and fill them in!

X	1	2	3	4	5	6	7	8	9	10
1										
2										
3										
4										
5										
6										
7										
8										
9										
10										

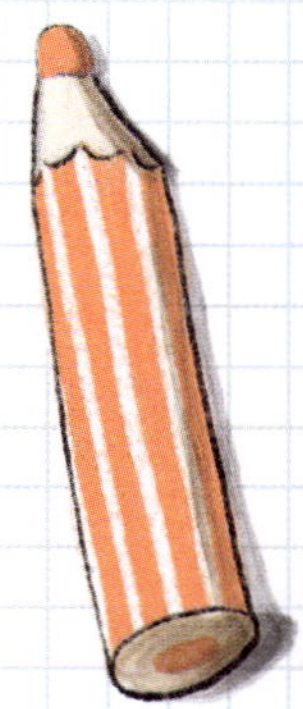

Ostomachion

It might sound like a strange sci-fi monster,
but an Ostomachion is actually a very old shape game.

Ancient idea

It was invented by Archimedes, an ancient Greek who lived around 2,200 years ago.

It's a square made up of 14 sections, and it looks like this

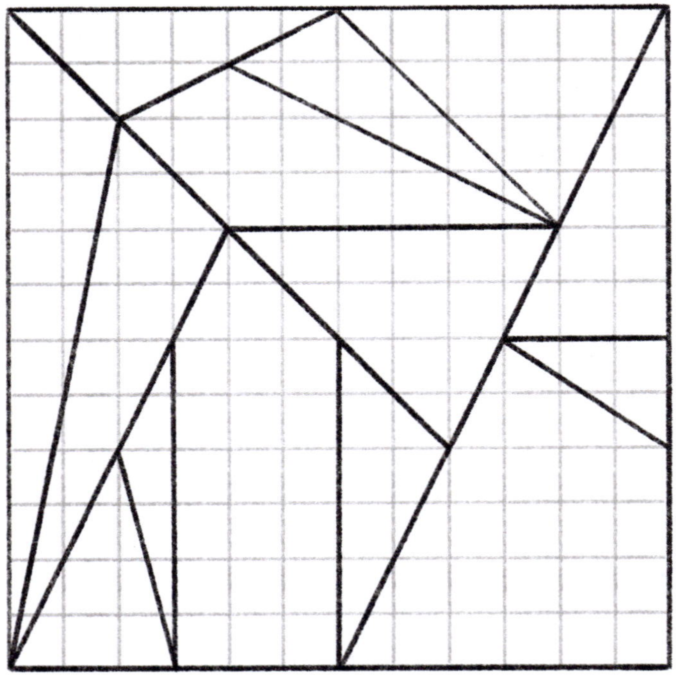

Archimedes found that you could rearrange the pieces in many different ways to make a square.

Try it!

To make your own ostomachion, trace the square and all the shapes, or copy them onto graph paper using a ruler and pen.

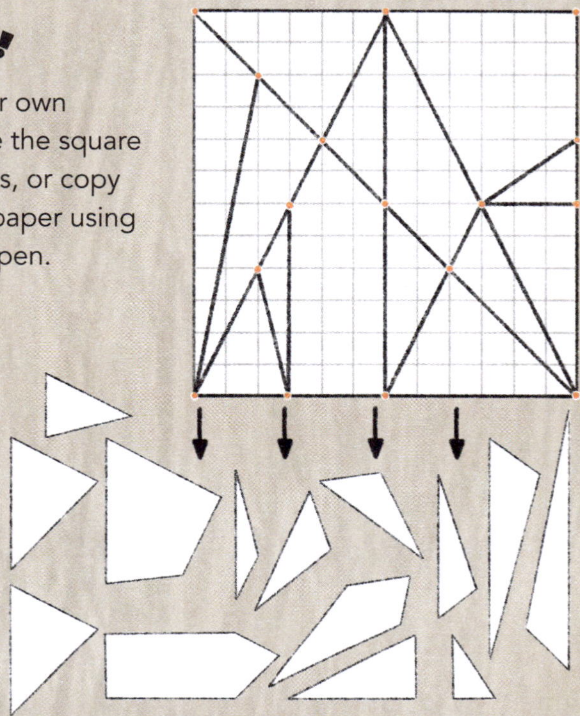

Carefully cut along all the lines, so that you have 14 pieces.

Try and fit them together in a square again.

See page 94 for the answers.

Create your own

You don't have to stick with Archimedes' version.

Use this empty square to design your own ostomachion

Then copy or trace it onto card and cut the pieces out.

Endless road tiles

Make these tiles, and however you fit them together, the roads will join up!

How does it work?

You just need white or light card, a ruler, a pencil, pens, and scissors.

Measure, draw, and cut out 9 card squares, each 10 cm x 10 cm (4 inches x 4 inches).

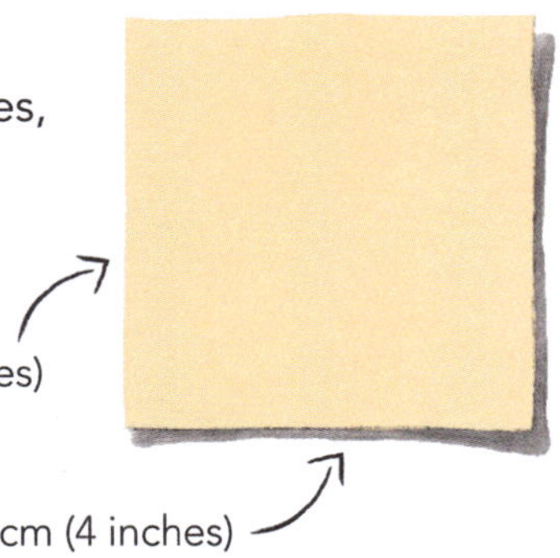

Make a pencil mark half way along the side of each square at the 5-cm (or 2-inch) point, like this.

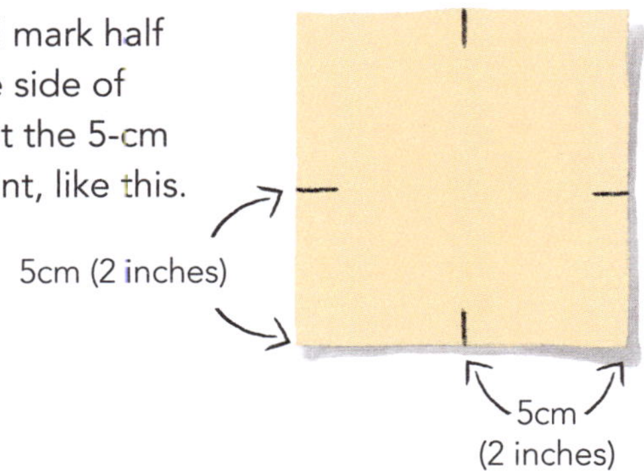

Add the roads

Now add the roads! On each card, draw any roads you like. You just have to make sure they all join the edges of the card where the marks are. For example …

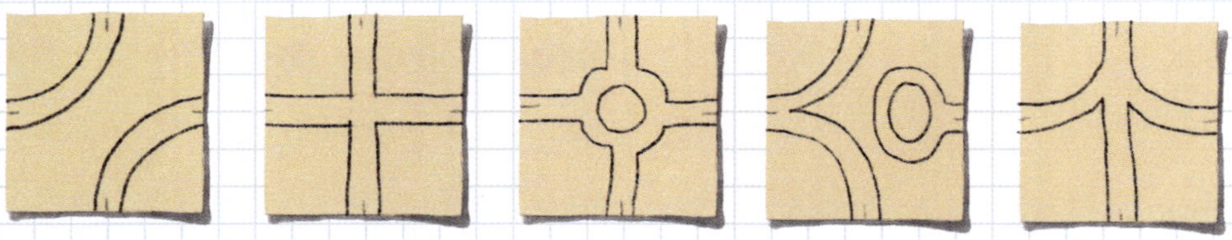

Now try fitting them together.

However you arrange them (and however many you make), the roads will link up.

Toy car traffic

You could make a set of bigger cards, perfect for toy cars ...

Harder cards!

What if you put TWO marks on each side of the squares?

You'll get even more options—like these!

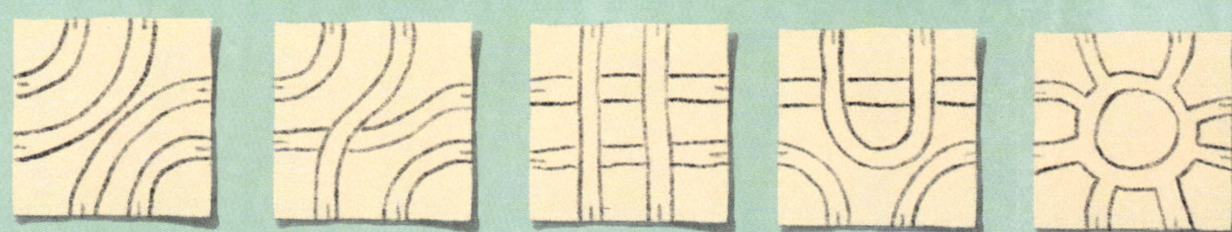

Abacus-cadabra!

How did people add and subtract before calculators and computers? They used an abacus!

What is it?

An abacus is a frame with rows of beads on it, which can move from side to side.

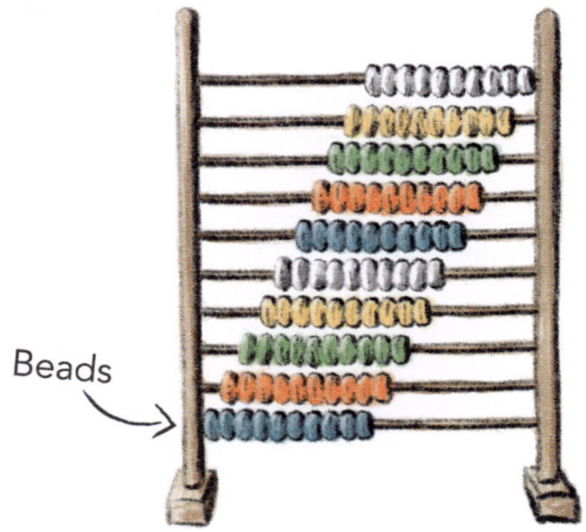

Ancient abacus

Abacuses have been around for thousands of years.

Here's one from Ancient Rome!

Try it yourself

You can use an abacus to show numbers, and to add numbers up.

You might have an abacus, because they're often sold as toys.

If not, you can use ours!

You just need some game counters, buttons, or small circles cut out of paper or card.

Arrange them along the lines, leaving space for them to move.

How it works

On an abacus, the top row of beads stands for single numbers, or 1s.
The next row stands for 10s.
The next row is 100s.
And so on!

So to make 234, for example, you would move these beads to the right.

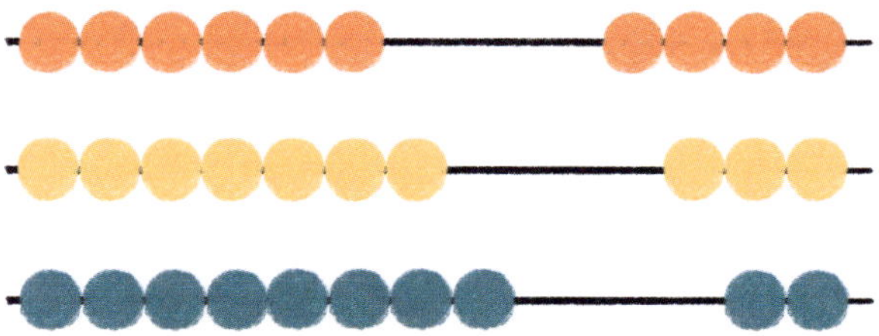

4	1s	4
3	10s	30
2	100s	200

You can also add up. To add another number, push it over to join the first one.

234 plus 215

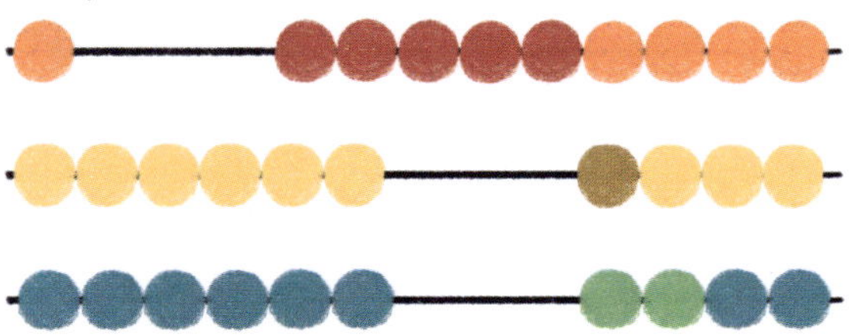

9	1s	9
4	10s	40
4	100s	400

Makes 449!

Abacus numbers

Can you draw these numbers on the abacuses by adding beads?

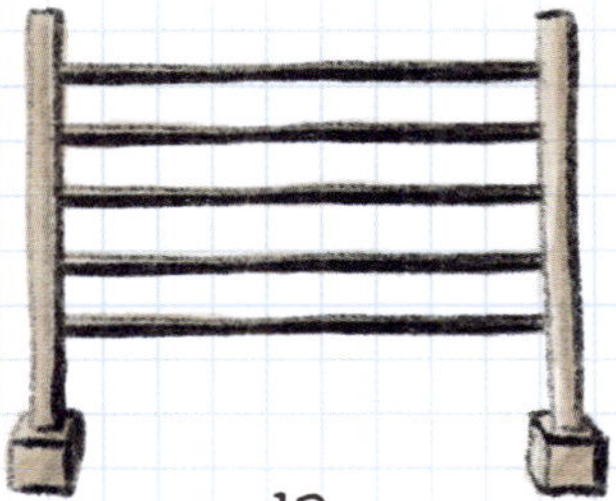

12

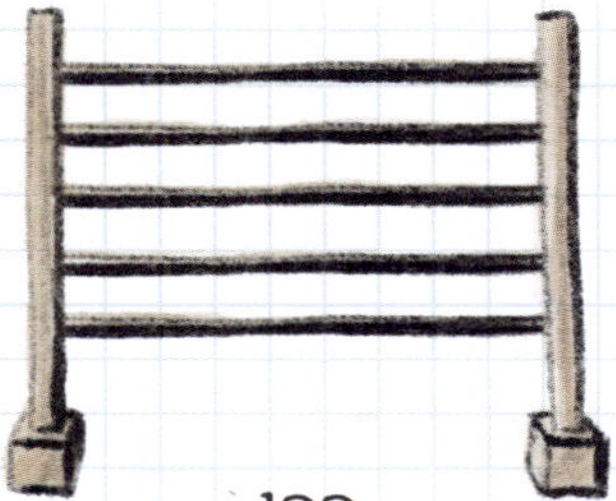

199

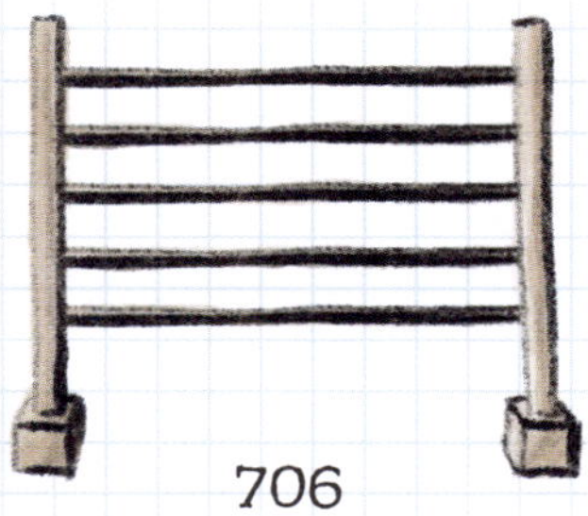

706

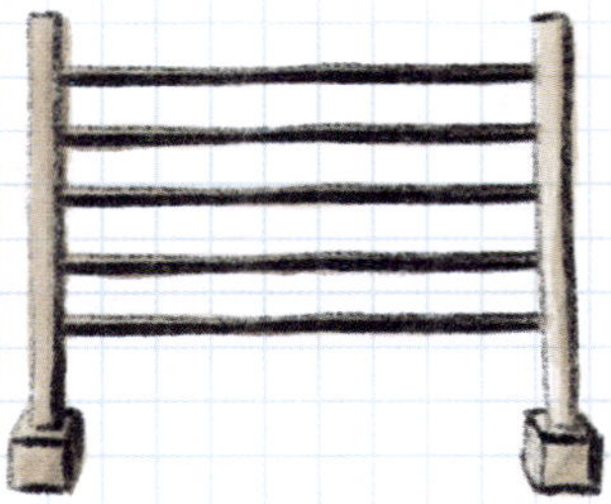

1,281

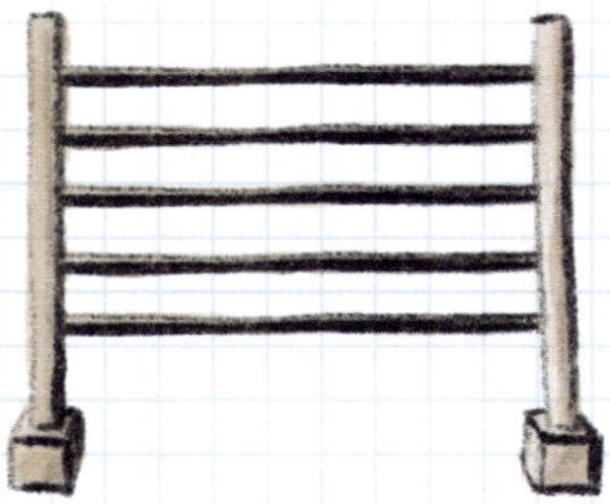

9,999

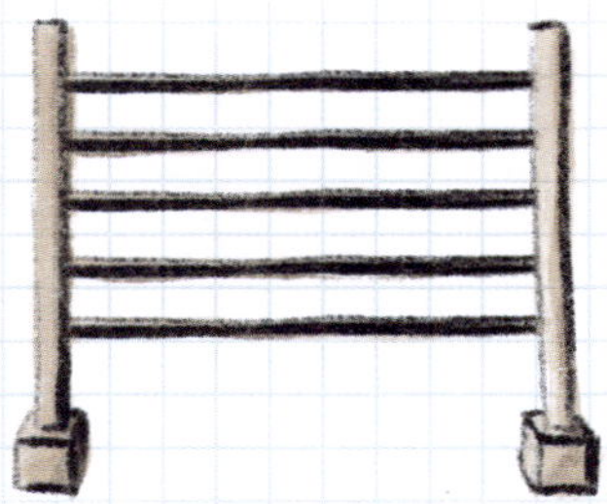

20,534

Maze magic

Amazing maze methods and magical maze know-how!

Many mazes

Mazes work in a variety of ways.
See if you can solve all these ...

In this kind, you have to get to the middle.

Huge hedge mazes often work like this.

In this ancient maze design, you have to go along every path to get to the end.

This type of maze has an entrance and an exit on opposite sides.

You often see these in puzzle books.

Maze cracker

For the last type of maze, there's a simple maze-cracking method.

All you have to do is stick to the right! (Or the left—either will work.)

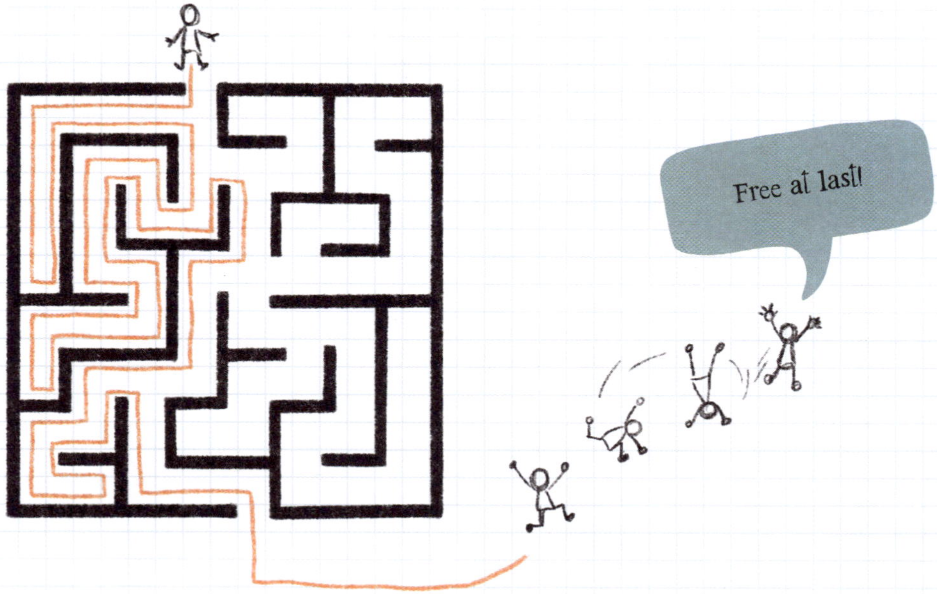

Enter the maze, then turn right whenever you can. You'll go along the side of the wall until you're out!

Make a maze

Now design your own maze for someone else to solve!

Here's a grid to help you plan.

Draw the correct route first.

Then draw the rest of the maze around it, and erase the solution at the end!

Maps and plans

A map shows you a place viewed from above. It looks much smaller than in real life!

Plan your space!

A plan of a room or outside space is the same.

Real outside space

Plan

It shows a bird's-eye view, but much smaller

It's all about the scale!

Maps and plans work using a "scale"—meaning how much smaller everything is!

For example, if the scale is 1 to 100, or 1:100, the map or plan is 100 times smaller than real life.

So, if this park is 100 m (100 yards) across …

This 1:100 park map is 1 m (1 yard) across

All the objects, details, and paths in the map or plan are 100 times smaller too

Try it yourself!

Try making your own plan of a room, such as a bedroom or classroom. You'll need a tape measure, ruler, and pencil.

Measure the length of the walls, and write the lengths down.

Then draw a plan of the room here.

For every 1 m (1 yard) in real life, use 10 squares.

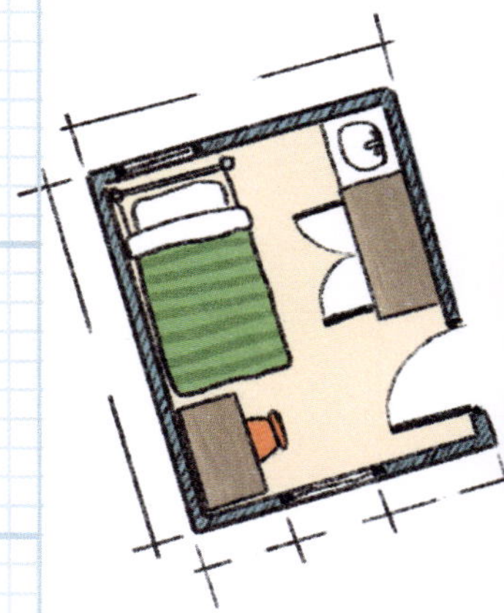

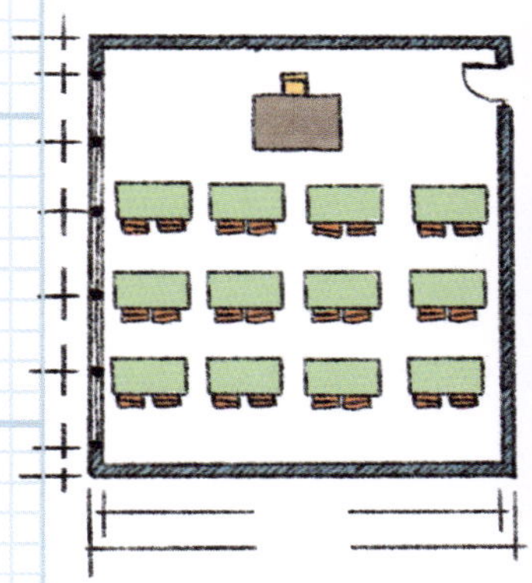

Impossible routes

Boggle your mind with this brain-teasing bridge challenge ...

The Seven Bridges

In the 1700s, people living in the city of Konigsberg used to wonder if it was possible to take a take a trip around their city, crossing all of its seven bridges only once.

Well, was it? See if you can do it!

How did you do? Impossible, right?

In 1736, mathematician Leonard Euler showed it couldn't be done.
He turned the problem into a simple diagram, like this:

Here are the four areas of land, and the bridges connecting them.

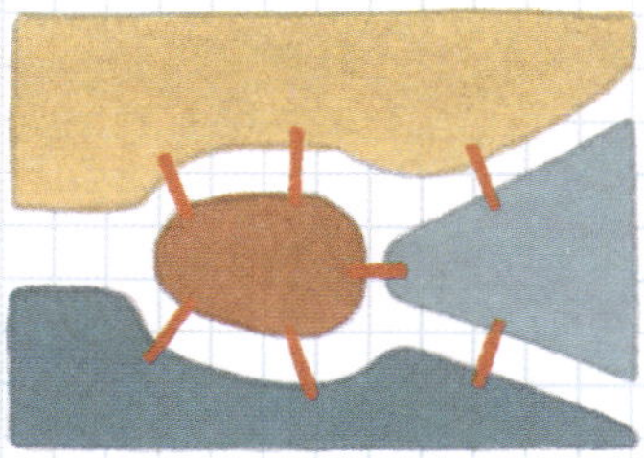

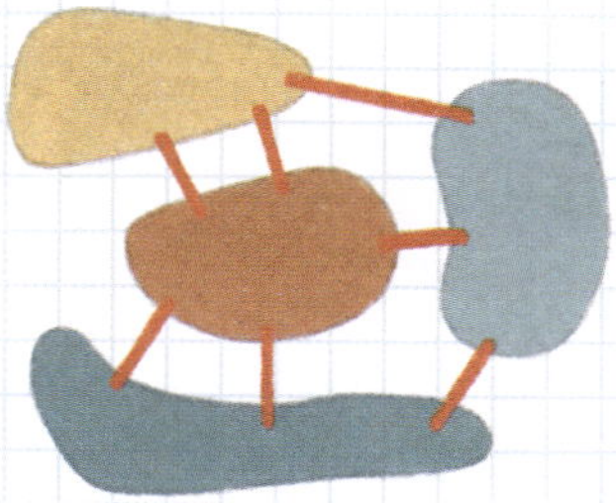

It doesn't matter what shapes the areas of land are, as long as the right number of bridges connect them.

You can even move the land and bridges around—see?

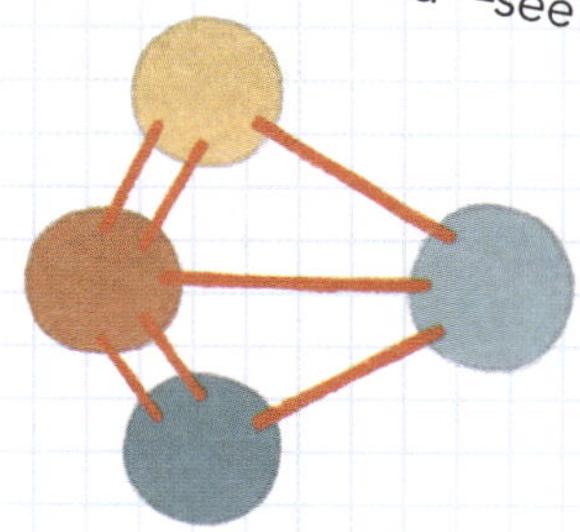

Here is Euler's diagram

Try drawing the same shape with a pencil, without lifting the pencil up or going over the same line twice. Bet you can't!

How about these?

Some are possible, some aren't!

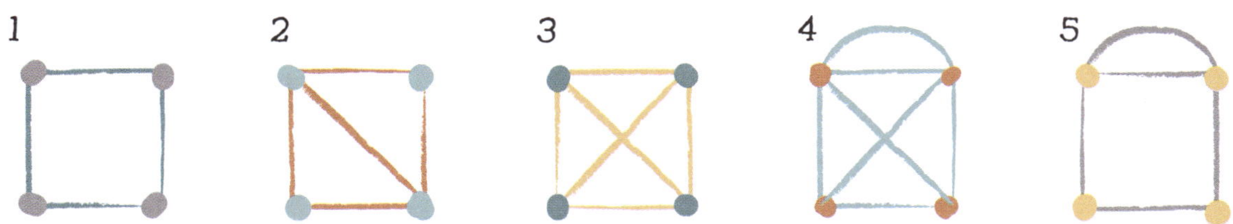

See page 94 for the answers.

Impossible Shapes

Did you know you can draw things that are **impossible**?

Here's how!

The artist M. C. Escher was always drawing impossible mathematical p ctures. His most famous ones involved impossible shapes and stairs. See if you can figure these out!

This impossible cube couldn't really exist in real life, but it can exist as a 3D picture that tricks your eyes.

Follow the edges of the cuboid with your eyes, or run your finger along them to see what happens!

Can you see how it works?

Escher's art often showed impossible steps, too, a bit like these ones. Can you see how the picture works?

Can you imagine walking around the path? You'd always go downstairs, never up!

Now it's your turn!

To start with something slightly easier, try drawing an impossible triangle.

Start with this triangle, which has parts sticking out.

Add these lines …

… then these lines …

And finally, the corners!

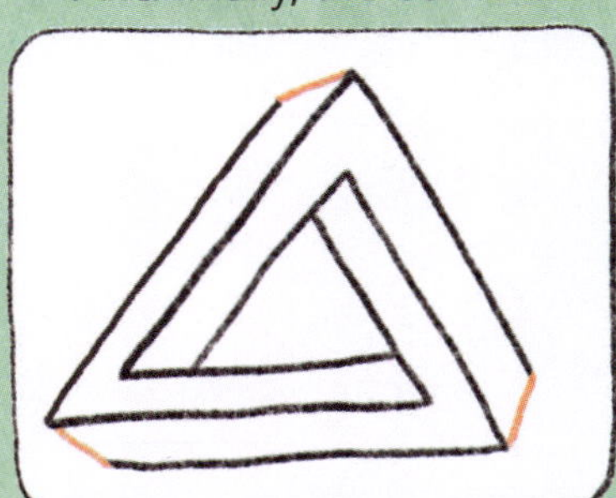

Now that you're an expert impossible artist, try copying this cube …

… and then draw the impossible fork!

Impossible loop

Unlike an impossible shape drawing, this loopy loop really does exist in real life!

One becomes ... one!

When you cut something in half, you get two halves ... right? Not always! When you cut this in half, it will stay in one piece.

Try it! You need a strip of paper like this.

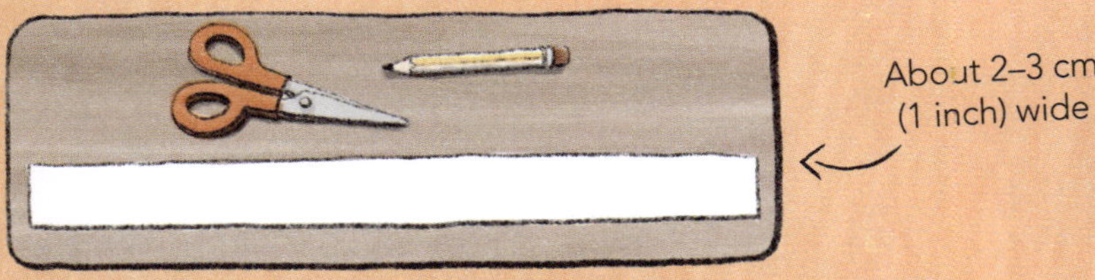

Make it into a loop. Flip one end over.

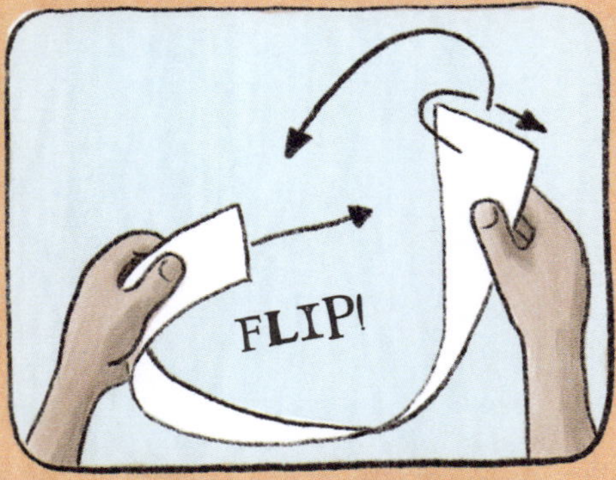

Glue or tape the ends together.

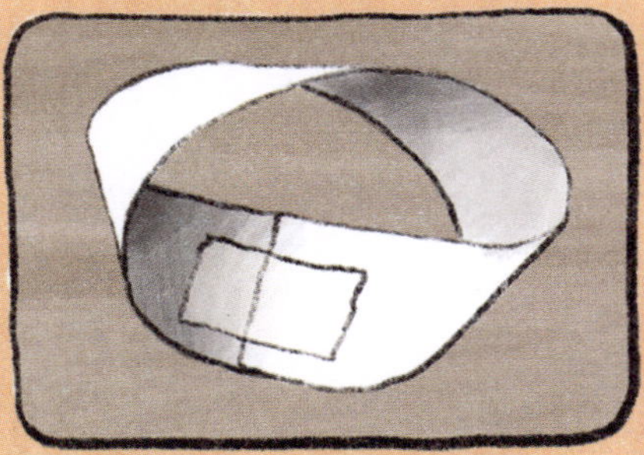

Now use pointed scissors to cut into the middle of the strip.

Then cut along the middle until you get back to where you started. You've cut it in two!

Except ... you haven't!

What???

This is called a Mobius strip—a loop with a twist in it. Flipping the end over attaches the opposite sides of the strip together. When you cut it in half, they stay connected, making a bigger loop!

Endless path

Artist M. C. Escher drew a picture of a Mobius strip with ants marching around it.

Use these outlines to draw your own versions.

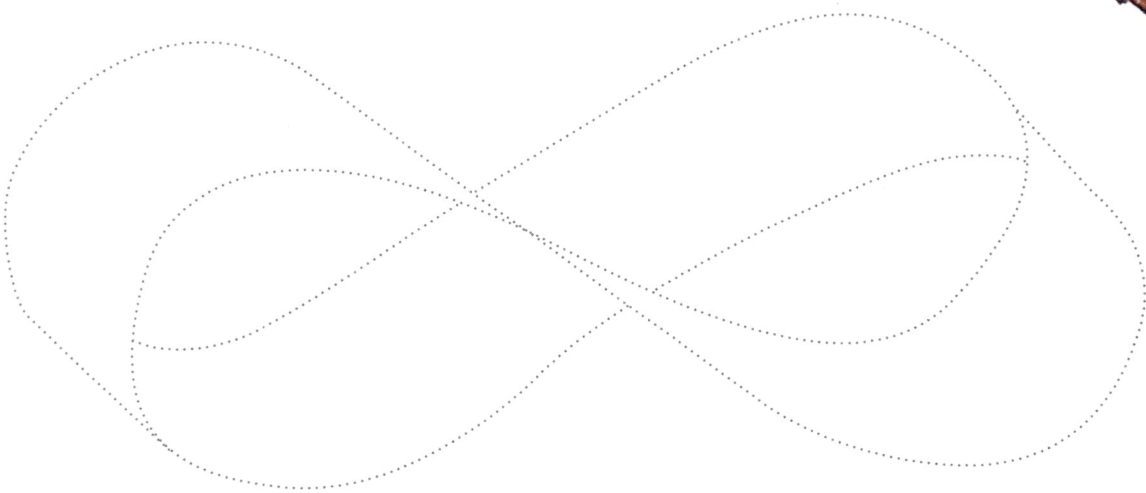

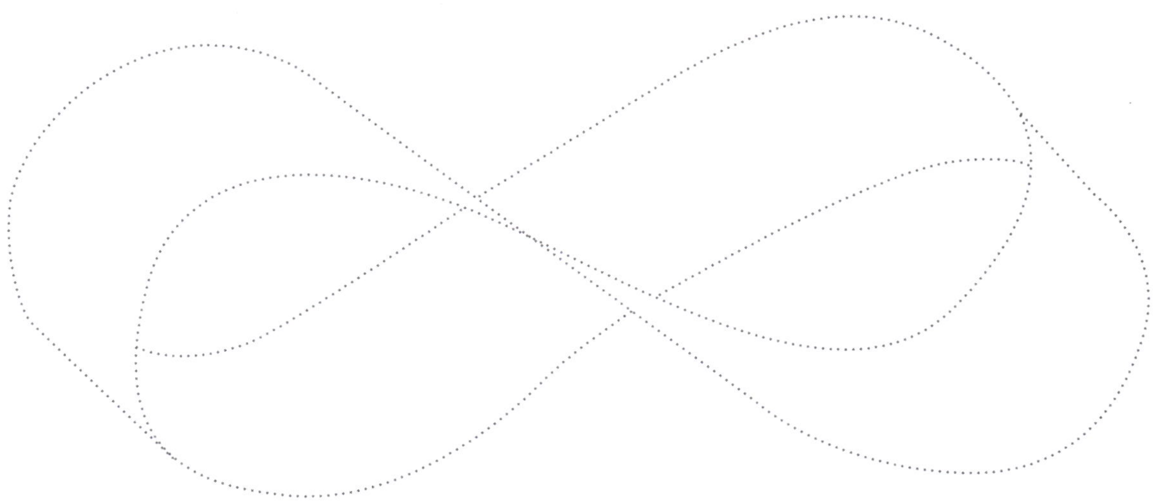

Googol it!

What do you think counts as a BIG number? A thousand? A million? A BILLION? What about a googol?

A what?

You've probably heard of Google, a search engine for finding things on the Internet. But "Google" was named after the googol, a truly ginormous number!

What is a googol?

To get to a googol, you have to know about using "power" with numbers. "Power" means how many times you multiply something by itself. For example ...

A **hundred** is 10 to the power of 2.

That means 10 x 10.

And we write is as a 1 with 2 zeros after it.

100

1 2

A **thousand** is 10 to the power of 3.

That means 10 x 10 x 10.

And we write is as a 1 with 3 zeros after it.

1000

1 2 3

So what's a googol?
A googol is much, MUCH bigger!

It's 10 to the power of 100.

That means:

10 x 10 x 10 x 10 x 10 x 10 x 10 x 10 x 10 x 10
10 x 10 x 10 x 10 x 10 x 10 x 10 x 10 x 10 x 10
10 x 10 x 10 x 10 x 10 x 10 x 10 x 10 x 10 x 10
10 x 10 x 10 x 10 x 10 x 10 x 10 x 10 x 10 x 10
10 x 10 x 10 x 10 x 10 x 10 x 10 x 10 x 10 x 10
10 x 10 x 10 x 10 x 10 x 10 x 10 x 10 x 10 x 10
10 x 10 x 10 x 10 x 10 x 10 x 10 x 10 x 10 x 10
10 x 10 x 10 x 10 x 10 x 10 x 10 x 10 x 10 x 10
10 x 10 x 10 x 10 x 10 x 10 x 10 x 10 x 10 x 10
10 x 10 x 10 x 10 x 10 x 10 x 10 x 10 x 10 x 10

Write it down!

So that means we write it as 1 with 100 zeros after it!

Write down a googol here! → 1

How big is that?

A googol is more than the number of grains of sand in the whole world. It's more than the number of stars in the universe.

Disappearing tunnel

With just a few circles, you can create a cool optical illusion that looks like a tunnel, disappearing into this book!

Simple circles

You'll need a compass so you can draw accurate circles of different sizes, along with a pencil, a ruler, and a black pen.

Start by drawing a basic circle. Set your compass to 5 cm (2 inches), so that your circle will be 10 cm (4 inches) across.

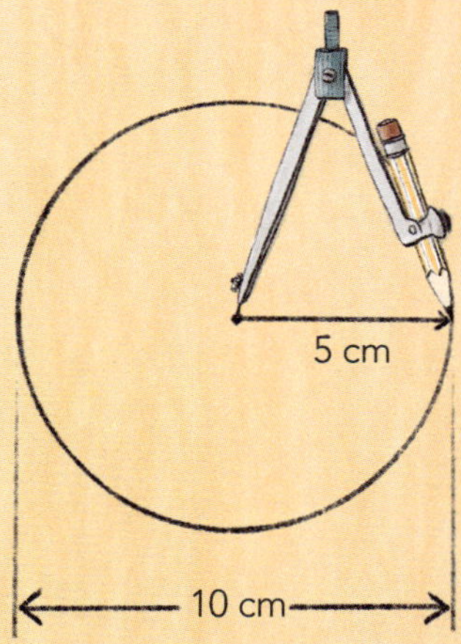

Draw a small dot at the bottom edge of the circle.

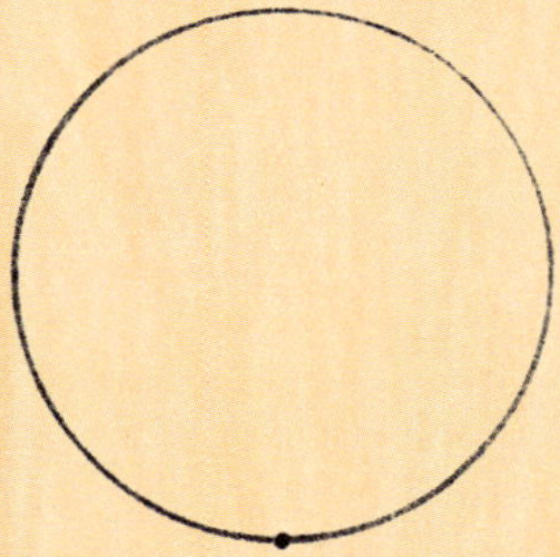

Now make your compass ½ cm (¼ inch) narrower, and draw another circle inside the first one, making sure the edge touches the dot—like this:

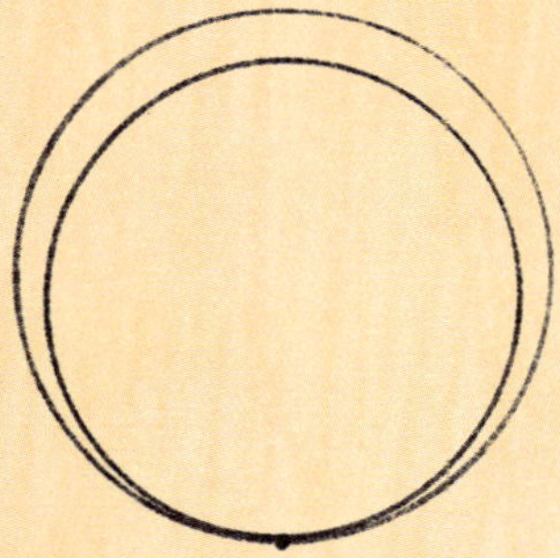

Keep doing this until the circle is full:

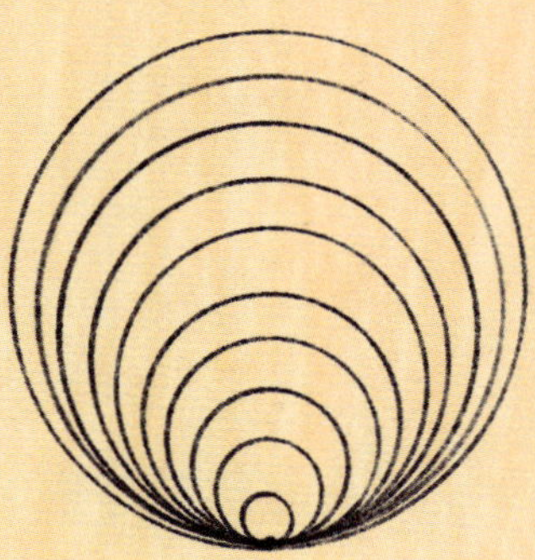

Fill in every other ring in black, like this:

Take a look!

Now look at the book from the bottom edge of the page to see the tunnel effect!

Try these!

Can you copy these versions, too?

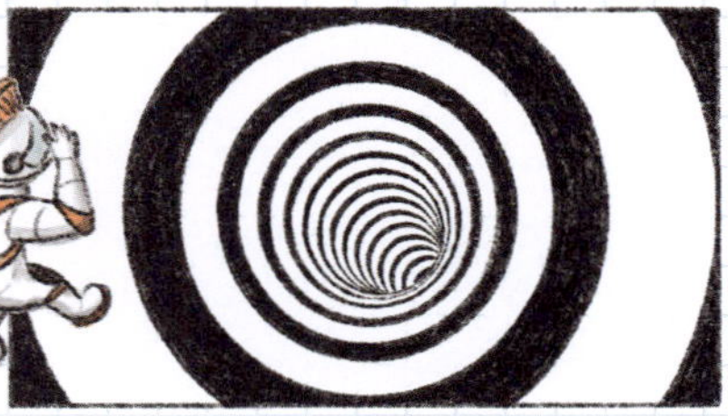

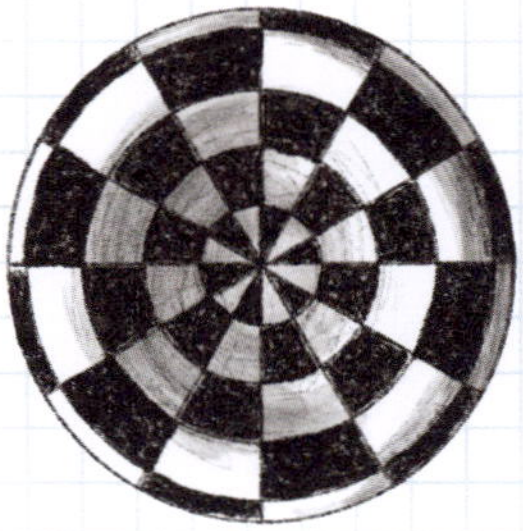

Floating ball

Here's another amazing 3D-effect drawing —this time it's floating in midair!

Stretch it out!

When you see a shape drawn on paper from a low angle, it looks shorter.

So, to make a realistic floating ball, you need to draw a long oval shape.

You need a pencil and a ruler.

Near the top of the page, measure and mark four small, faint lines, like this.

Use them to help you draw a smooth oval:

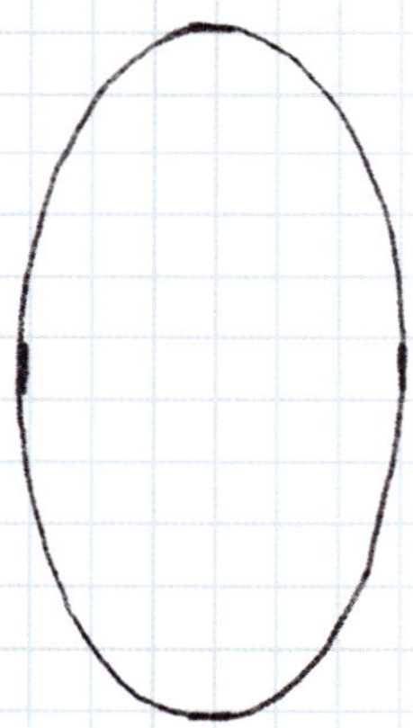

Carefully shade in your oval, leaving a white area near the top, and blend to darken at the bottom …

Add a shadow, like this

Now look at your drawing from the bottom of the page, at a low angle …

Ta-da! It's floating!

Cut it out!

For an even more realistic effect, draw your oval on a piece of paper, then cut off the top, like this:

Try these!

Can you make a floating cone or cube, too?

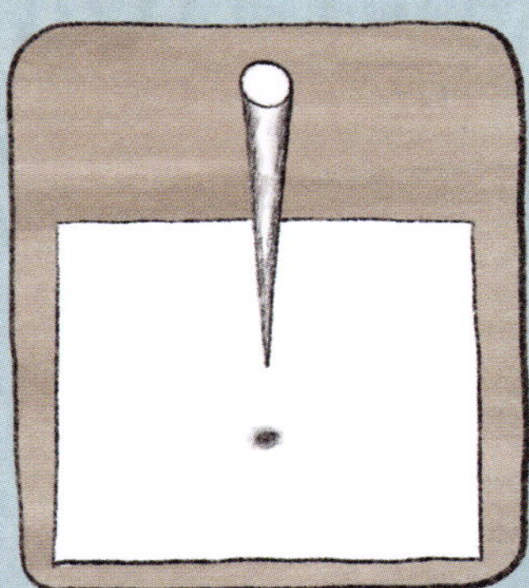

They should look like this on the paper:

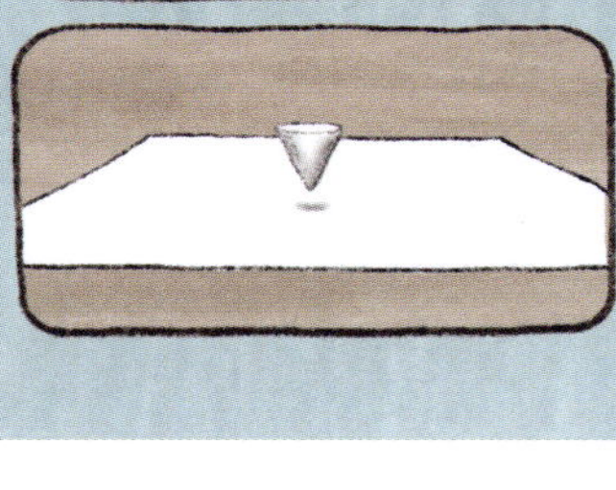

Magic magnifier

What if you want to copy a picture, but make it much bigger?
You need some mathematical magic!

Try this!

To try it out, start with this rabbit.

First, take the picture you want to copy, and draw a grid of squares over it using a pencil and ruler. Measure the lines so that they are all the same distance apart (1 cm, or ½ inch, is a good size).

That's me!

If you can't draw on the picture, draw your grid on tracing paper and lay it over the top!

Then, to make a bigger copy, draw a grid of bigger squares on a piece of paper.

For example, you could make all the lines 2 cm (1 inch) apart.

We've given you one here to start with!

To copy the picture, look at each square, and carefully copy the line into the bigger square, making it look as similar as possible—like this

Did it work?

When you've copied all the squares, you should have a whole bigger picture!

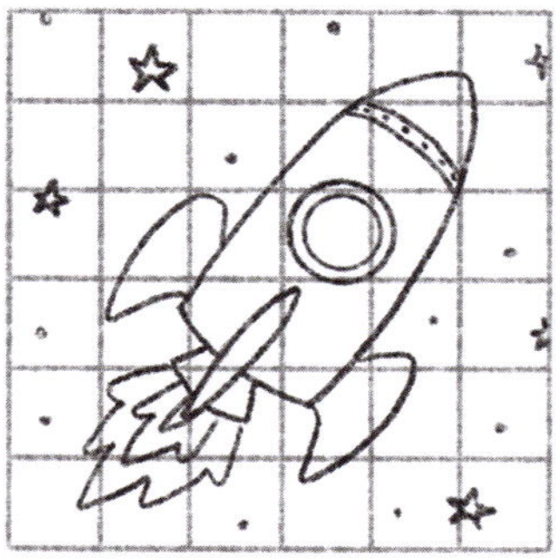

Magnify one of our pictures, or find one of your own to use.

Cool codes

Secret agents need to send secret messages ... and for that, they need secret codes!

Code basics

In a coded message, each letter is replaced by a different letter or symbol.
Here's a basic symbol code.

A	B	C	D	E	F	G	H	I	J	K	L	M	N	O	P	Q	R	S	T	U	V	W	X	Y	Z
@	!	&	*	%	<	>	?	€	#	∞	§	Σ	^	★	∂	Ω	≈	£	¶	✪	¿	+	≠	/	≥

Σ % % ¶ Σ % @ ¶ Σ € * ^ € > ? ¶

.... (Answer on page 94!)

Code-makers often use numbers and mathematical shapes to create codes.

Letter grid

Here's one method.
The letters are arranged in a 5 x 5 square grid.

(That means you have to leave one out!
We've left out Q—just use K instead.)
To write a message, you choose a rule,
such as "Right 1."

Then change each letter to the letter one square to the right of it.

So A becomes B

A	B	C	D	E
F	G	H	I	J
K	L	M	N	O
P	R	S	T	U
V	W	X	Y	Z

For J, you go back around and use F

Different rules make different codes—such as ...

Down 1

Right 2

Down 3

Secret messages

To crack the code, use the grid at the bottom of page 76. Reverse the rule to find each letter.

Try these!

RULE: Right 1:

PO NZ XBZ

....

RULE: Down 2:

LBTYR MSUMZWKDO

....

Pigpen code

Here's another one to try.

You put the alphabet letters in grids with dots, like this.

Then copy the shape for each letter.

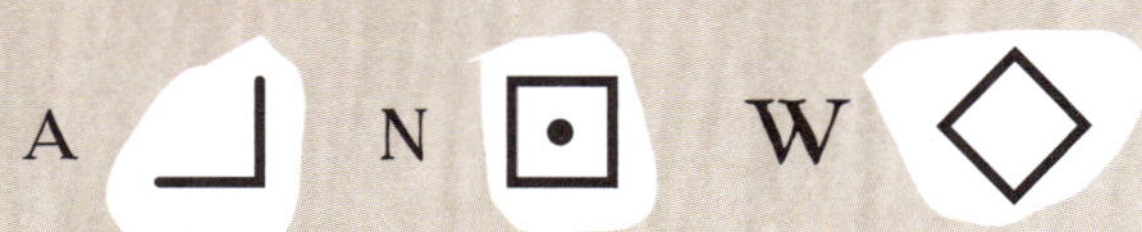

....

See page 94 for the answers.

Bigger and bigger

When you double a number, you multiply it by 2 and get twice as many. But what happens if you keep doing that?

Folding challenge

For this folding challenge, you need a large piece of paper. A sheet of newspaper or old wrapping paper is perfect.

Fold it in half. You've doubled it, so 1 layer has become two

Now fold the folded sheet in half again. Ta-da! You now have FOUR layers

How many times do you think you can keep folding—and how many layers will there be?

As you'll discover, it soon gets very difficult!

That's because when you double a number again and again, it soon grows to a very big number. This is called exponential growth.

Exponential earwigs

Here's another way to look at it.

Imagine that you start with one earwig …

… who has two babies …

… then each baby has two babies …

… and each of them has two babies

And so on!

How many rows would you need to get to 100 babies?

Take a guess, then fill them in to find out!

Body numbers

Did you know that your body is great at mathematics?
Check these measurements, and see for yourself ...

Face and hand

Your hand is about the same length as your face, from hairline to chin:

Hand = Face!

Of course, we're all different, but these measurements are often amazingly accurate!

If they don't work for you, see if you can find other measurements that do match.

Foot and forearm

For most people, the length of your foot matches the length of your forearm.

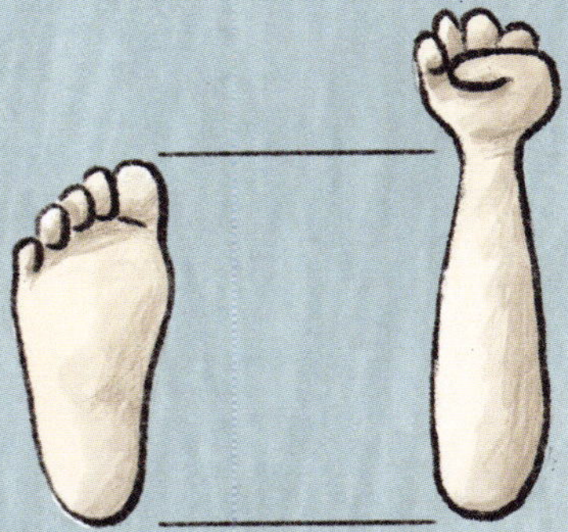

Foot = Forearm!

Height and arm span

Most people's height matches the distance between their fingertips when they spread their arms out:

Height = Arm span!

Body units

Ever wondered why the "foot" unit of measurement is called a foot?

That's right—because it's based on a foot! (An adult man's foot, to be precise).

Today, the foot has a standard length of 30 cm (12 inches).

1 foot

1 foot

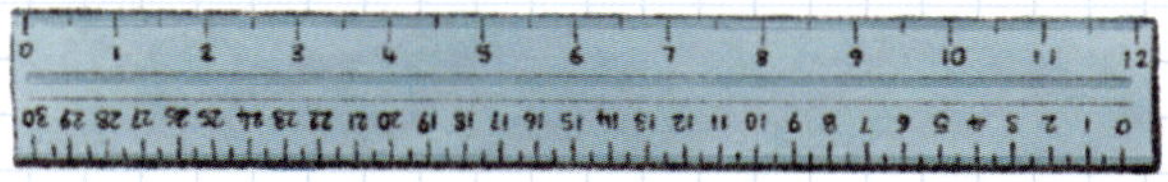

These old measurements are based on the human body, too:

Hands

used to measure horses

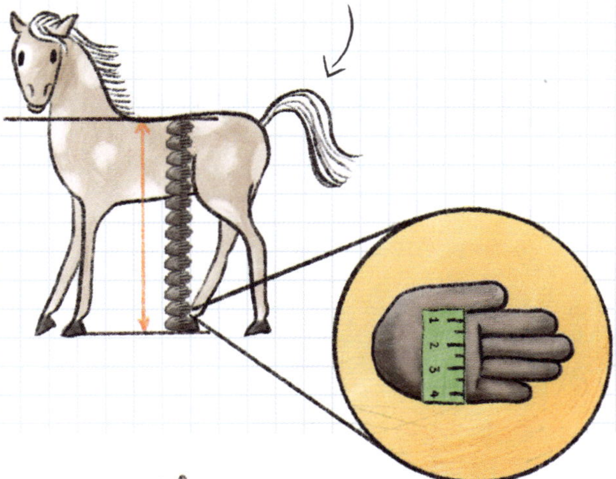

Inches

based on the width of a man's thumb

A fathom

is 1.8 m (6 feet), based on a man's arm span

Your height in hands

Can you measure your own height in your own hands?

I am hands tall!

Mathematical nature

Mathematics doesn't just exist in books like this one, or in school. It's everywhere in nature!

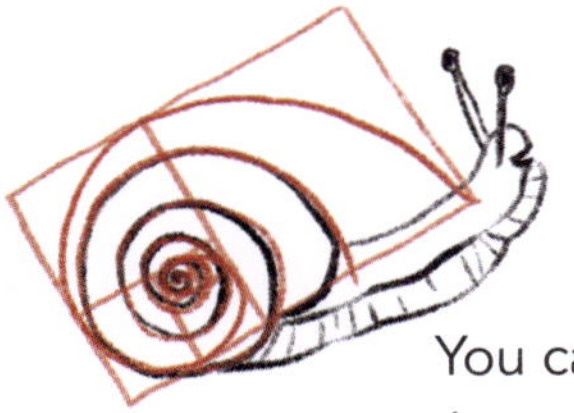

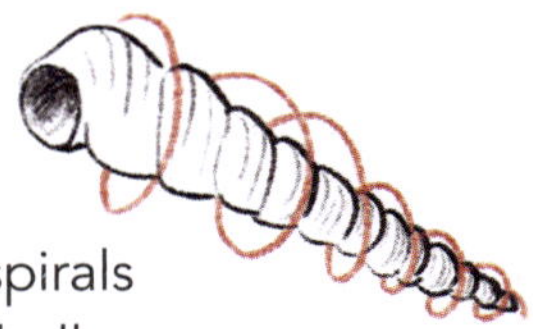

Spirals

You can find beautiful spirals in snail shells and seashells.

Some flowers grow in a spiral pattern, like this rose.

So do some pine cones and animal horns …

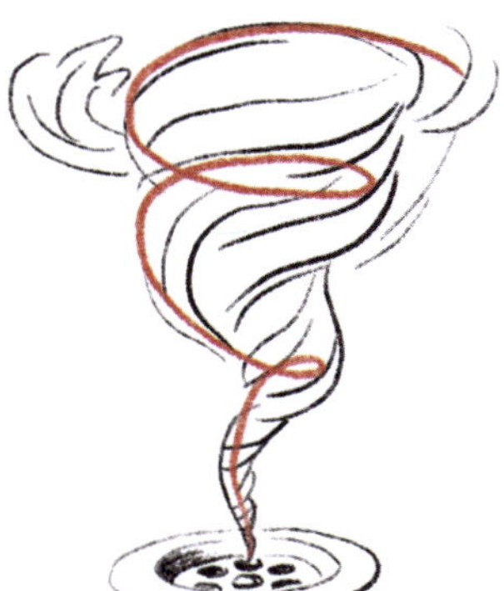

There's a spiral in this spider's web …

Circles

Can nature make a perfect circle? Think of bubbles, an egg yolk, ripples in a pond, or the iris of an eye!

Symmetrical shapes

Animals and plants often have symmetrical shapes and parts (see page 22) …

And what about snowflakes?

Tessellating shapes

Besides honeybees' hexagon combs (see page 49), there are other tessellating patterns in nature:

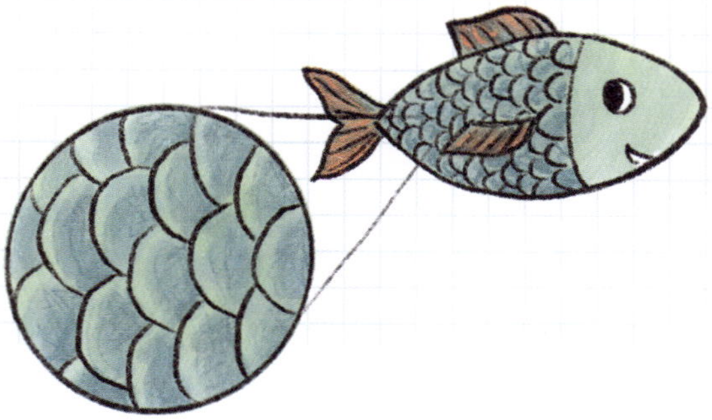

Scales on fish, snakes, and lizards

The skin of a pineapple

Pattern-spotting

Look for patterns in nature next time you're in a garden or park, on a walk, or at the seaside. Or look at pictures of plants, animals, rocks, water, and other natural things in books, magazines, or on the Internet.

Sketch what you find here!

A spiral

Tessellating shapes

Symmetrical shapes

A perfect circle

Something else interesting!

Mirror magic

One way to think of symmetrical shapes (see page 22) is that one side is a mirror image of the other.

This means that if you put a mirror next to anything, you can make a symmetrical shape.

You can get small mirrors like this from a toy store, tile store, or makeup store.

You might already have some in your home or school.

Crazy reflections

You can make crazy faces by using a mirror to reflect part of your face, like this:

Next time you are out shopping, use a shiny window to reflect half your body, like this!

Two mirrors

If you have two small mirrors, you can make an even cooler effect.

Hold them at an angle, like this …

… with a picture, pattern, object or drawing here

Reflect this!

Try using your mirror or mirrors on these pictures and patterns …

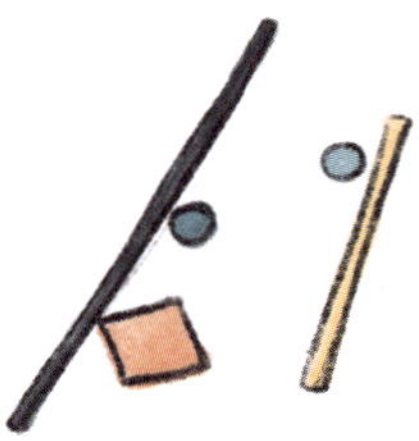

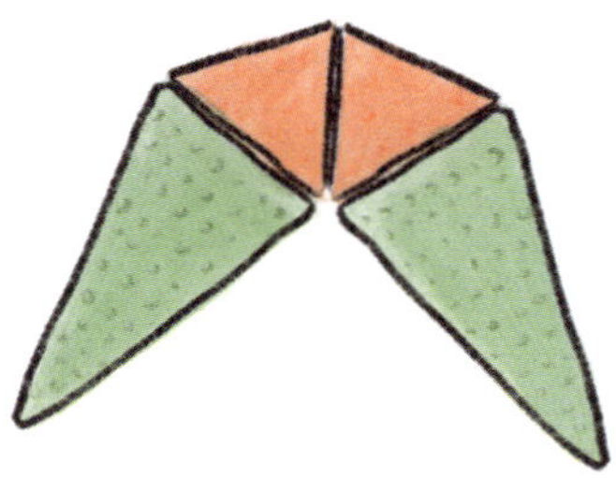

100%!

Percent, written as "%," means "out of 100."

So 100% means 100 out of a 100, or a complete whole.

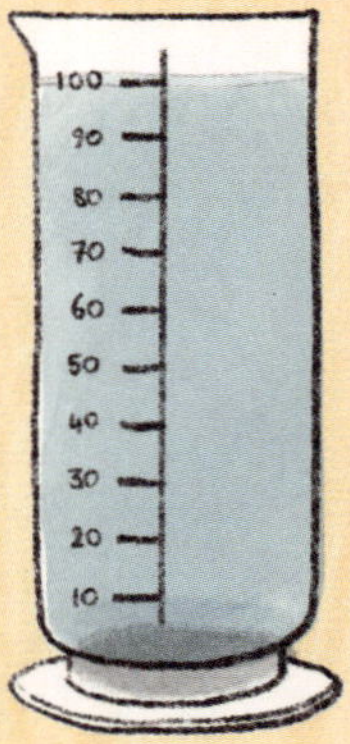

50% means 50 out of a 100, which is a half, or ½.

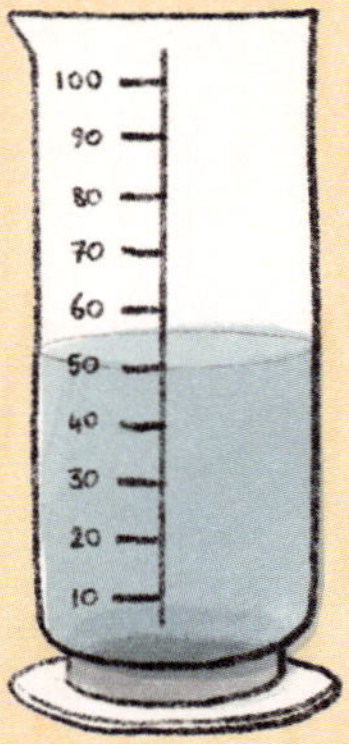

20% means 20 out of a 100, which is a fifth, or ⅕.

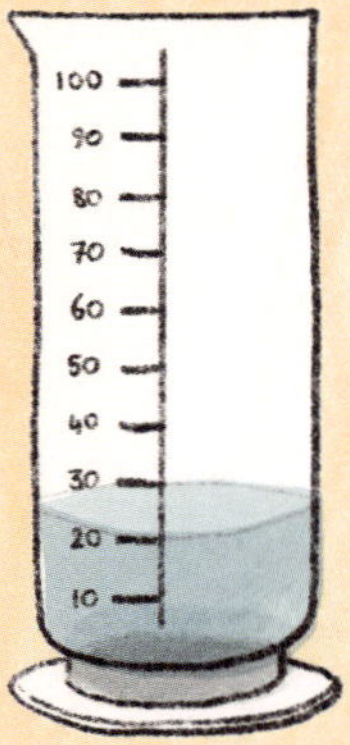

You can have a percentage of anything.

So, for example, 50% of 20 is 10.

The percentage trick

If a percentage is hard to figure out, here's a handy tip: Percentages work both ways around!

So, for example:
6% of 50 …

Is the same as 50% of 6!

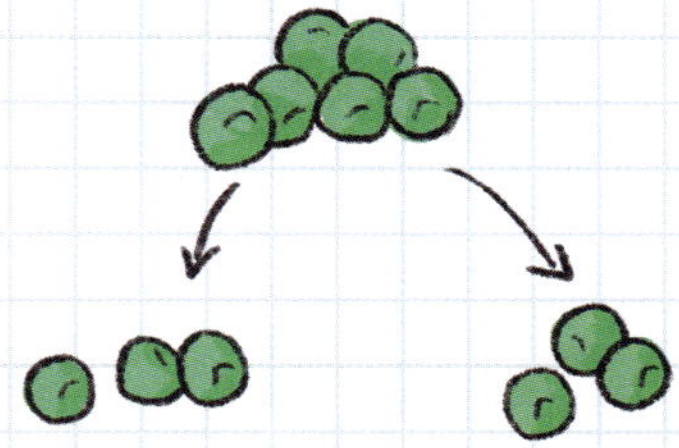

That's half of 6 … or 3. Simple!

Extra every time

Workers are building a new tower. Every month it gets 20% higher.

Can you fill in the numbers and draw the tower for the next month after that?
(Remember, for the next step, you need 20% of 120, not 100.)

Right now, it's 100 m tall. (You can think of the tower as 100 yards tall, if that's easier for you.)

Next month, it will be 20% taller.

20% of 100 is … 20

Adding that to 100 makes … 120

So it will be …

120 m tall.

20% of 120 is … ____

Adding that to 120 makes … ____

So it will be …

____ m tall.

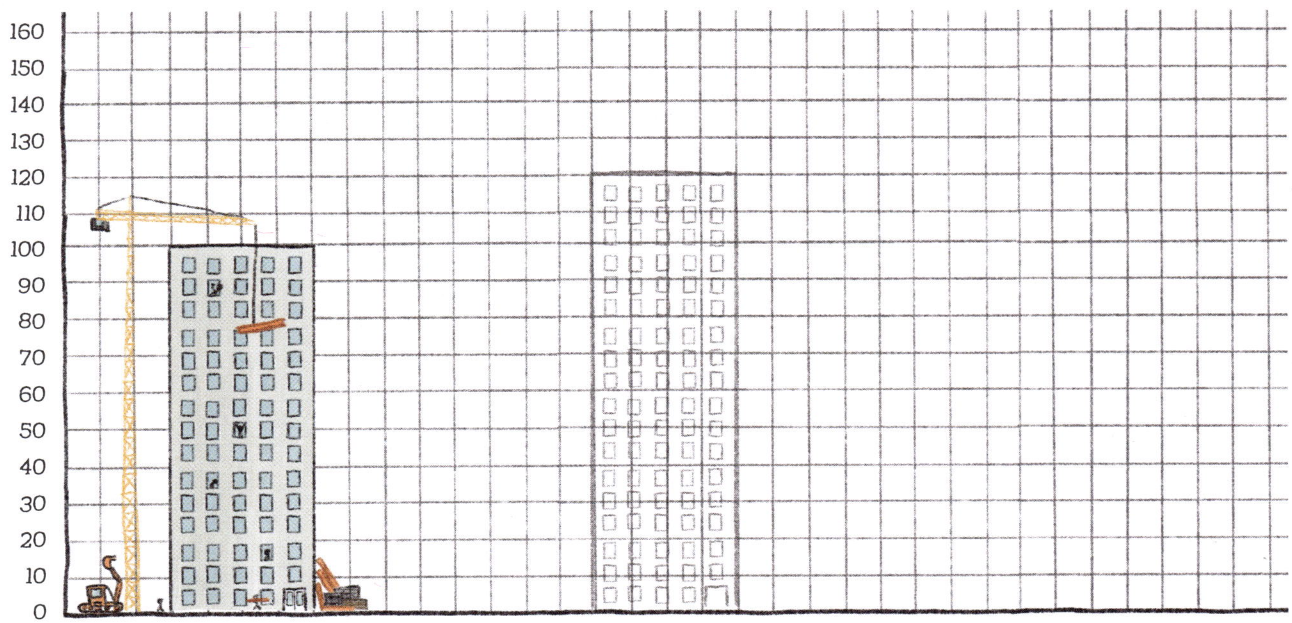

Answer on page 94!

Fool your brain!

How do optical illusions work?
For some of them, it's mathematical!

Your brain constantly does calculations to help you understand what you're seeing.

Illusions trick you by making your brain get its numbers wrong!

The strange sticks

Is one line bigger than the other?
Which one?

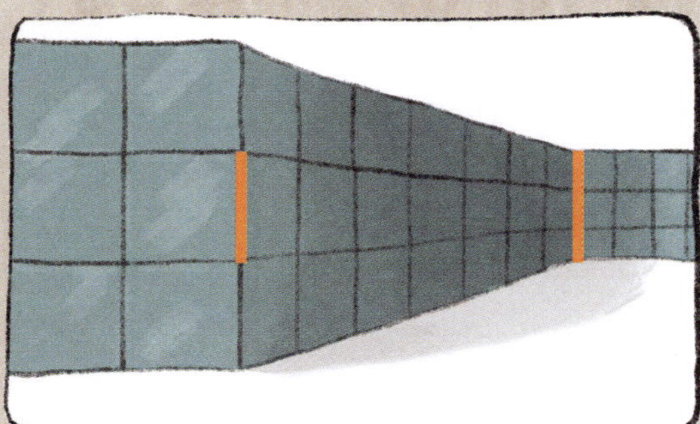

ARE YOU SURE!?

Grab a ruler and measure them!

They're the same length! But your brain follows the blue squares and thinks the second line must be farther away. Faraway things usually look smaller. So your brain decides that the faraway stick must be really big!

The curious curves

Now try this one.
Which shape is bigger?

They're the same—again! Trace one, and it will fit over the other.

Your brain wrongly compares the two side-by-side curves, sees that one is longer, and gets confused.

Now try this!

Would you believe it if we said that these two tables are exactly the same shape? You might think one is long and thin, and the other is short and thick. But would you be right?

You'll have to check them to find out!

What to do

First, get some tracing paper or parchment paper, and carefully trace the shape of the first tabletop.

Cut it out, then turn the shape sideways. See if it fits over the other table. If it's an exact fit, they ARE the same shape!

Huh? How does it work?

This amazing illusion works because we're used to seeing tables from the side or the end, not from directly above. When you see a table, your brain figures out how long and wide it must be by taking into account how far into the distance it reaches. So your brain thinks that the first table must be really long, and the second one must be really wide.

Add some legs, and you have two matching tables!

Or DO you?

Below zero!

You know all about how numbers work, right? They start at zero, then they just keep going up ...

Negative numbers

But did you know there are lots more numbers on the other side of zero?

They're called the negative numbers, or minus numbers.

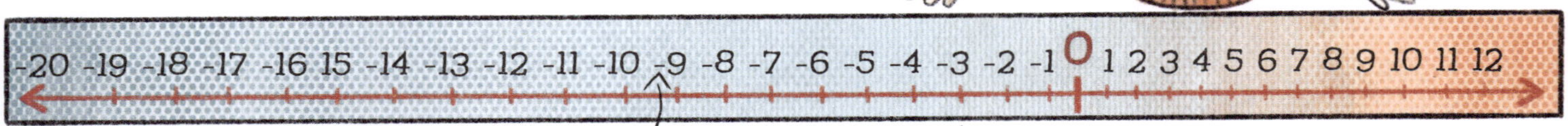

The minus sign here shows that it's negative!

For example, -3 is 3 below zero

The numbers above zero are called positive numbers

Add and subtract

You can do sums with negative numbers, too.

For example, what's 4 take away 7?

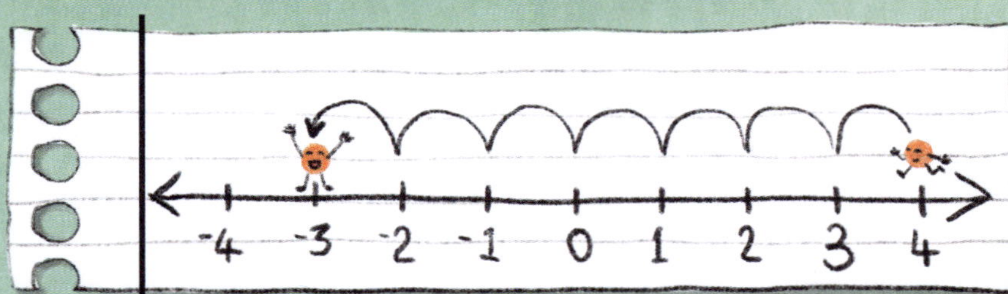

You just count back 7 places from 4, and you get to –3!

Brrrrrr!

Negative numbers have all kinds of handy uses in everyday life.

For example, what happens when it gets really, REALLY cold?

The temperature drops below zero!

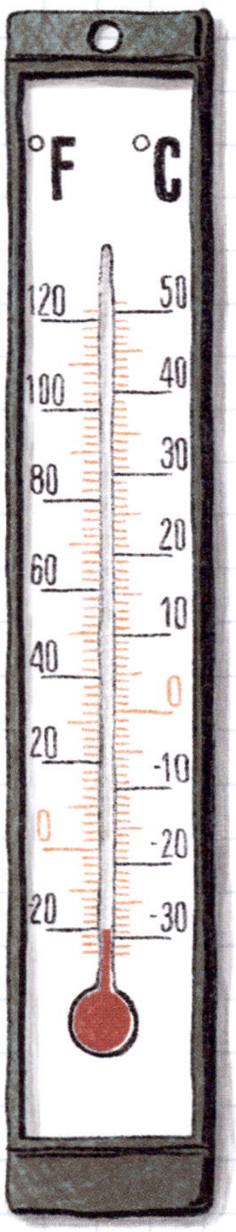

Whether you're using Fahrenheit (F) or Celsius (C), the minus numbers continue down below zero as it gets colder

Counting down ...

Can you figure out what the temperatures will fall to, then mark them on the thermometers?

The temperature will fall by 20 degrees tonight!

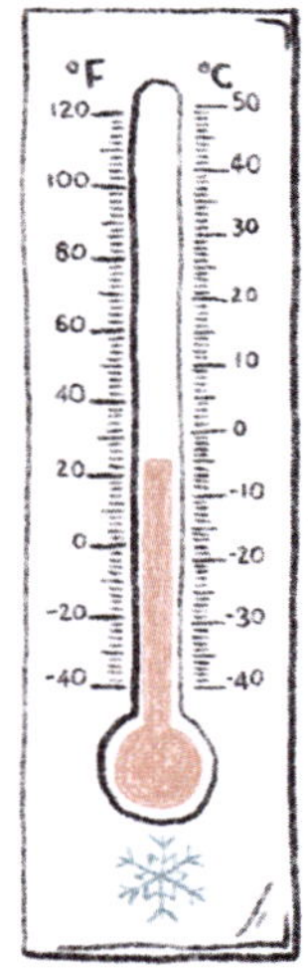

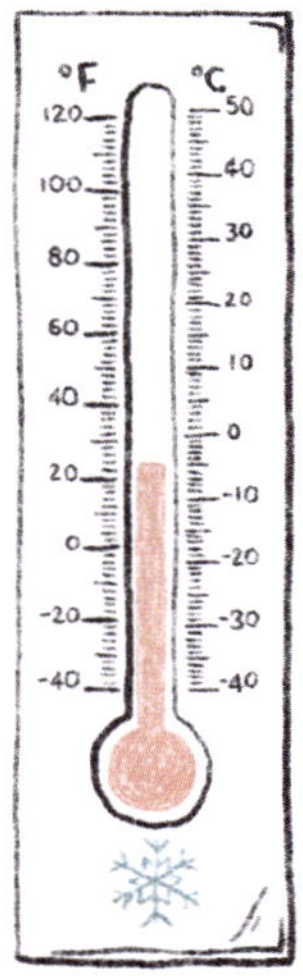

The temperature will fall by 10 degrees tomorrow!

The temperature will fall by 25 degrees on Tuesday! Wear a coat!

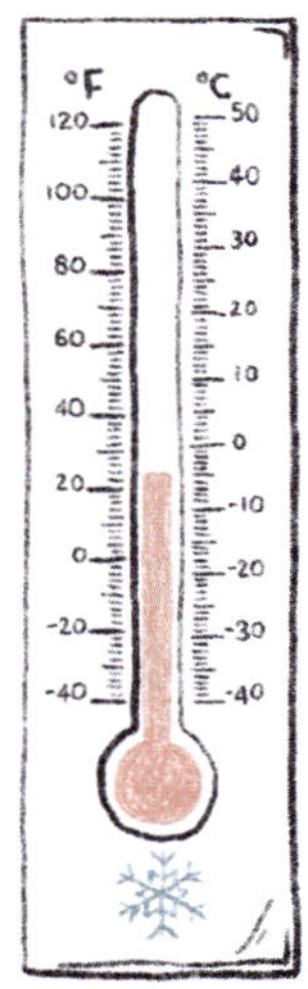

Forever and ever and ever ...

Have you ever wondered just how many numbers there are? The answer is, numbers go on for ever! They're countless—or "infinite."

Endless numbers

"Infinity" means endlessness. Numbers go on forever.

How do we know? Because whatever the biggest number you can think of is, you could always add 1!

Or you could multiply it by 10!

Or multiply it by a googol (see page 68)!

Smaller and smaller parts

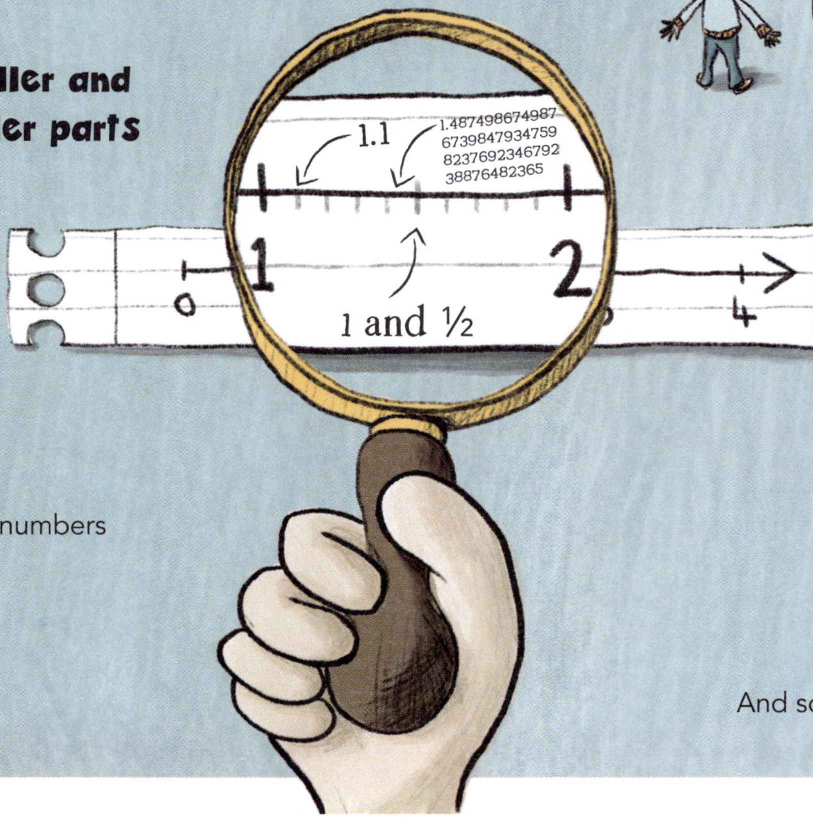

And there's more. Thanks to fractions and decimal numbers, we can divide any number into an infinite number of smaller parts.

So there's an infinite number of numbers just between 1 and 2!

And so on!

Hui Shi's stick

According to legend, ancient Chinese thinker Hui Shi said:

"If you take a stick and cut off half of it each day,
it will last forever."

Is that true?

Here's a stick.

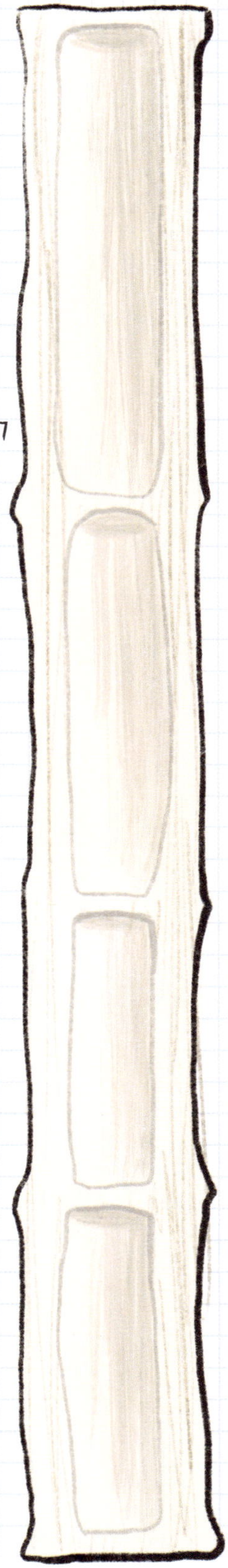

Draw a line dividing it in two.

Then draw a line dividing one of the halves in two.

And keep going! What happens?

If you had a pen thin enough, you could keep dividing it forever!

Answers

Page 17
The green rope is not a knot.

Page 19
5	1	6
6	1	4
5	2	3

Page 39

Page 41
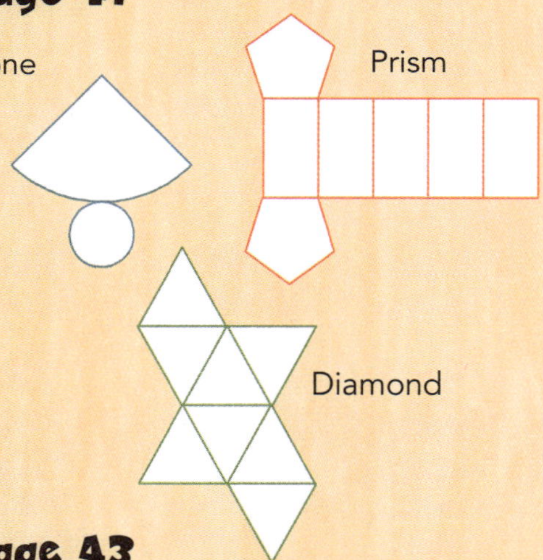

Page 43
Triangle trick 1: (But there could be other ways, too!)

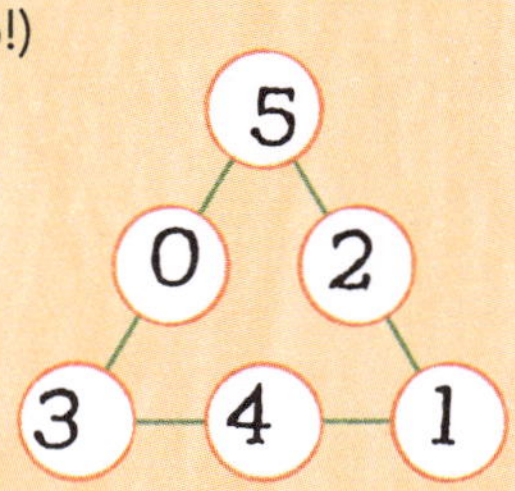

Triangle trick 3: Answer: 27

Page 47

Farmer Fallow's fence
Answer: 10m

Page 53
Possible solutions:

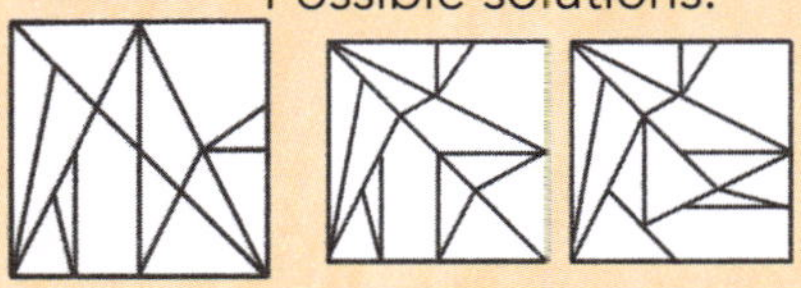

Page 63

1,2, and 5 are possible. 3 and 4 are not possible.

Page 76–77

Cool codes
∑ ⁒ ⁒ ¶ ∑ ⁒ @ ¶ ∑ € * ^ € > ? ¶
Answer: MEET ME AT MIDNIGHT

PO NZ XBZ
Answer: ON MY WAY

LBTYR MSUMZWKDO
Answer: BRING CHOCOLATE

Answer: I AM OVER HERE

Page 87

Extra every time
Answer: 20% of 120 is 24
Adding that to 120 makes 144
So it will be 144 m tall

Glossary

3D Short for three-dimensional, meaning a real solid object, or a picture that looks like one.

Abacus A frame with rows of beads on it that can move from side to side, used for counting and calculating.

Axle A rod or bar that fits through the middle of a wheel.

Circumference The distance all the way around the edge of a circle.

Code A set of letters or other symbols used to make hidden messages.

Compass A tool for drawing a circle (sometimes called a pair of compasses).

Cone A 3D shape with a flat circular base and straight sides sloping into to a point.

Coordinates Numbers or letters used to mark squares or lines on a grid, used to pinpoint a particular location.

Cube A 3D shape with six square sides.

Cuboid A 3D shape with six sides that are all rectangles.

Cylinder A 3D shape with two circles or ovals at the ends and straight sides.

Decimal number A number with a decimal point with numbers after it that show a fraction, such as 1.5.

Degree A measurement of temperature; or a measurement of one -360th of a circle.

Diameter The length across the middle of a circle.

Dice A cube with different numbers of dots on each side, used for playing games (also called a die).

Equilateral triangle A triangle in which all the sides are the same length.

Fathom A unit of measurement about 1.8 m (6 feet) long, based on a man's arm span.

Foot A unit of measurement about 30 cm (12 inches) long, based on the length of a man's foot.

Fractal A mathematical pattern that repeats itself.

Fraction A part of a number, such as a half, quarter, or tenth.

Googol A very large number that is written as a 1 with 100 zeros after it.

Grid A set of lines crossing each other at regular intervals.

Hand A unit of measurement about 10 cm (4 inches) long, based on the width of a hand.

Hexagon A flat shape with six sides.

Honeycomb A structure bees build out of wax, made up of hexagon-shaped chambers.

Inch A unit of measurement based on the width of a man's thumb.

Infinite Endless or countless.

Isosceles triangle A triangle in which two of the sides are the same length.

Knot A tangle in a string that cannot be untied by pulling the ends.

Mobius strip A loop-shaped strip with a half-twist in it.

Negative numbers or minus numbers Numbers below zero, written with a minus sign in front of them, such as -3.

Net A flat shape that can be folded up to make a 3D shape such as a cube or pyramid.

Number sequence A sequence of numbers that follows a pattern or set of rules.

Optical illusion A picture that fools the brain into seeing something that isn't there, or seeing something differently from how it really is.

Oval A flat shape that looks like an egg or a flattened circle (also called an ellipse).

Ostomachion A shape puzzle made by cutting a square into several smaller shapes.

Percent (written as "%") A part of 100—so, for example, 25% means 25 out of every 100, or a quarter.

Pi The number you get if you divide a circle's circumference by its diameter: about 3.14.

Pixels (short for picture elements) Tiny squares or other shapes used to make up a picture.

Prism A 3D shape with two matching shapes opposite each other and straight sides.

Pyramid A 3D shape with a flat base and triangular sloping sides that meet in a point at the top.

Radius The distance between the middle and the edge of a circle.

Reuleaux triangle A type of triangle with curved sides that is the same width across the middle whichever way you measure it.

Right angle A 90-degree angle, like the corner of a square.

Right-angled triangle A triangle in which one of the corners is a right angle, like the corner of a square.

Scale The size of a map or plan compared to the size of the thing it shows, such as 1:100.

Scalene triangle A triangle in which all the sides are different lengths.

Sierpinski triangle A type of fractal pattern made up of triangles.

Spiral A pattern made by a line that curves outward from a central point.

Square To multiply a number by itself; or a flat shape with four equal sides and four right-angled corners.

Squared or graph paper Paper covered with a grid of crossing lines that make squares.

Square number A number that is another number squared (or multiplied by itself).

Symmetrical A symmetrical shape is made up of matching, mirror-image parts on both sides.

Tessellate If a shape can tessellate, it can fit together to fill a space with no gaps.

Tetrahedron A 3-sided pyramid with a triangular base and three triangle sides.

Unknot A tangle in a string that can be untied by pulling the ends.